Everyday Mathematics®

The University of Chicago School Mathematics Project

Home Links

Grade **1**

Wright Group

The **McGraw·Hill** Companies

The University of Chicago School Mathematics Project (UCSMP)

Max Bell, Director, UCSMP Elementary Materials Component; Director, *Everyday Mathematics* First Edition; James McBride, Director, *Everyday Mathematics* Second Edition; Andy Isaacs, Director, *Everyday Mathematics* Third Edition; Amy Dillard, Associate Director, *Everyday Mathematics* Third Edition

Authors

Jean Bell, Max Bell, John Bretzlauf, Amy Dillard, Robert Hartfield, Andy Isaacs, James McBride, Rachel Malpass McCall*, Kathleen Pitvorec, Peter Saecker

*Third Edition only

Technical Art	**Teachers in Residence**	**Editorial Assistant**
Diana Barrie	Jeanine O'Nan Brownell	Rossita Fernando
	Andrea Cocke	
	Brooke A. North	

Contributors

Robert Balfanz, Judith Busse, Mary Ellen Dairyko, Lynn Evans, James Flanders, Dorothy Freedman, Nancy Guile Goodsell, Pam Guastafeste, Nancy Hanvey, Murray Hozinsky, Deborah Arron Leslie, Sue Lindsley, Mariana Mardus, Carol Montag, Elizabeth Moore, Kate Morrison, William D. Pattison, Joan Pederson, Brenda Penix, June Ploen, Herb Price, Dannette Riehle, Ellen Ryan, Marie Schilling, Susan Sherrill, Patricia Smith, Robert Strang, Jaronda Strong, Kevin Sweeney, Sally Vongsathorn, Esther Weiss, Francine Williams, Michael Wilson, Izaak Wirzup

Photo Credits

©Brand X Pictures/Alamy, p. 47; ©Burke/Triolo Productions/Getty Images, p. 47; ©Frederic Cirou/Getty Images, p. 51; ©Ralph A. Clevenger/CORBIS, cover, *center*; ©Ken Davies/Masterfile, p. 245; ©FoodCollection/IndexStock, p. 167; ©Jules Frazier/Getty Images, p. 18; Getty Images, cover, *bottom left*; ©giantstep/Getty Images, p. 167; ©K.Hackenberg/Masterfile, p. 169; ©Ken Karp photography, p. 49; ©Tom and Dee Ann McCarthy/CORBIS, cover, *right*; ©Ryan McVay/Getty Images, pp. 47, 49; ©Siede Preis/Getty Images, p. 49; ©Stockdisc/Getty Images, p. 49; ©SW Productions/Getty Images, p. 2; ©ThinkStock LLC/IndexStock, p. 167.

www.WrightGroup.com

Wright Group

Send all inquiries to:
Wright Group/McGraw-Hill
P.O. Box 812960
Chicago, IL 60681

ISBN-13 978-0-07-609738-8
ISBN-10 0-07-609738-2

7 8 9 DBH 12 11 10 09 08

The McGraw·Hill Companies

Contents

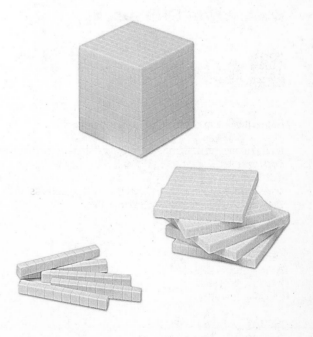

HOME LINK 1·1

Unit 1: Family Letter

Introduction to *First Grade Everyday Mathematics*

Welcome to *First Grade Everyday Mathematics.* It is part of an elementary school mathematics curriculum developed by the University of Chicago School Mathematics Project (UCSMP). *Everyday Mathematics* offers children a broad background in mathematics.

Several features of the program are described below to help familiarize you with *Everyday Mathematics.*

A problem-solving approach based on everyday situations By making connections between their own knowledge and experiences, children learn basic skills in meaningful contexts so that mathematics becomes "real."

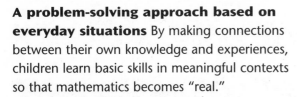

A variety of formats for frequent practice of basic skills Instead of practice presented only in a tedious drill format, children practice basic skills in a variety of engaging ways. In addition to completing daily mixed practice pages, finding patterns on the number grid, and working with addition and subtraction fact families, children will play games designed to develop basic skills.

An instructional approach that revisits concepts regularly To enhance the development of basic skills and concepts, children revisit previously learned concepts and practice skills encountered earlier. The lessons take advantage of previously learned concepts and skills and build on them throughout the year.

A curriculum that explores mathematical content beyond basic arithmetic Mathematics standards in the United States, as well as around the world, indicate that basic arithmetic skills are only the beginning of the mathematical knowledge children will need. In addition to basic arithmetic, *First Grade Everyday Mathematics* emphasizes the topics discussed on the following page.

◆ **Number and Numeration** Counting; reading and writing numbers; investigating place value of whole numbers; exploring fractions and money

◆ **Operations and Computation** Learning addition and subtraction facts, fact families, and extended facts; beginning informal work with properties of numbers and problem solving

◆ **Data and Chance** Collecting, organizing, and displaying data using tables, charts, and graphs

◆ **Measurement and Reference Frames** Using tools to measure length, capacity (quarts, liters), and weight; using clocks, calendars, timelines, thermometers, and ordinal numbers such as *fifth* and *tenth*

◆ **Geometry** Exploring 2-dimensional shapes (squares, triangles, rectangles) and 3-dimensional shapes (pyramids, cones, prisms)

◆ **Patterns, Functions, and Algebra** Exploring attributes, patterns, sequences, relations, and functions; finding missing numbers and rules in problems; studying properties of operations (addition and subtraction)

Everyday Mathematics will provide you with ample opportunities to monitor your child's progress and to participate in your child's mathematics experiences. Throughout the year, you will receive Family Letters to keep you informed of the mathematical content your child will be studying in each unit.

You will enjoy seeing your child's confidence and comprehension soar as he or she connects mathematics to everyday life.

We look forward to an exciting year!

2

HOME LINK 1·2

Unit 1: Family Letter

Unit 1: Establishing Routines

One purpose of this first unit is to help children become comfortable with a cooperative-learning environment in which they work together to build mathematical concepts. Another purpose is to introduce and establish routines that will be used this year and in the grades to come. This unit also reviews various mathematical concepts introduced in Kindergarten.

In Unit 1, children will review counting by 1s, 2s, 5s, and 10s. They will have opportunities to count and record numbers of various objects, such as hands, fingers, eyes, and ears. In addition, they will use pennies to count money, practice writing numbers, and begin to use a thermometer.

Vocabulary

Important terms in Unit 1:

Home Link A suggested follow-up or enrichment activity to be done at home. Each Home Link activity is identified by the following symbol:

tally A mark used in a count. Tallies let children represent numbers they can count and say, but cannot yet write.

$$\text{卌 /// is the tally count for the number 8.}$$

temperature How hot or cold something is relative to another object or as measured on a standardized scale such as degrees Celsius or degrees Fahrenheit.

tool kit A bag or box containing a calculator, measuring tools, and manipulatives often used by children in *Everyday Mathematics*.

3

Do-Anytime Activities

To work with your child on concepts taught in this unit, try these interesting and rewarding activities:

1. Count orally by 2s, 5s, and 10s when doing chores or riding in the car. Occasionally count down, or back; for example: 90, 80, 70, 60, ...

2. Take inventories around the house and while shopping. Have your child keep track of each count using tally marks.

   For example, count food items and nonfood items bought at the grocery store:

   +HT +HT +HT /   +HT //
   food items   nonfood items

3. Listen to and discuss weather reports with your child.

As You Help Your Child with Homework

As your child brings home assignments, you may want to go over the instructions together clarifying them as necessary. The answers listed below will guide you through the Home Links for Unit 1.

Home Link 1·9

1. Other possible answers include: TV listings, food packages (expiration dates), and clocks.

2. 1, 2, 3, 4, 5, 6

Home Link 1·10

1. Sample answer:

Number	Tally Marks
4	////
7	+HT //
12	+HT +HT //
16	+HT +HT +HT /
19	+HT +HT +HT ////

2. 1; 2; 4; 6; 8; 9

Home Link 1·11

1. Drawing should be of a Math Exploration.

2. 4   3. 7   4. 11

Home Link 1·12

1–2. Other possible answers include: oven, refrigerator, freezer, and thermostat.

3. 5   4. 3   5. 2

Home Link 1·13

1. Your child should draw a group of objects.

2. Sample number story: There are 5 flowers in the garden. If I pick 1 of them to give to my teacher, how many flowers will be left? Answer: 4 flowers

   NOTE: Encourage your child to come up with his or her own way to solve the problem, whether it's thinking logically, drawing pictures, or counting on fingers. As an adult, you know that $5 - 4 = 1$, but for your child, coming up with his or her own strategy is more natural than thinking of the number story as $5 - 4 = 1$.

4. 6   5. 9   6. 15

7. 1   8. 4   9. 10

HOME LINK 1·8

Numbers Are Everywhere

Family Note   Your child will bring home assignments called "Home Links." The assignments will not take much time to complete, but most of them involve interaction with an adult or an older child.

There are good reasons for including Home Links in the first-grade program:

◆ The assignments encourage children to take initiative and responsibility. As you respond with encouragement and assistance, you help your child build independence and self-confidence.

◆ Home Links reinforce newly learned skills and concepts. They provide thinking and practice time at each child's own pace.

◆ These assignments relate the mathematics your child is learning to the real world, which is very important in the *Everyday Mathematics* program.

◆ Home Links will give you a better idea of what mathematics your child is learning.

Listen and respond to your child's comments about mathematics. Point out ways in which you use numbers (time, TV channels, page numbers, telephone numbers, bus routes, shopping lists, and so on). *Everyday Mathematics* supports the belief that children who have someone do math with them, learn math. Fun counting and thinking games that you and your child play together are very helpful for such learning.

For this first Home Link, your child might look for a newspaper ad for grocery items, a calendar page, or a picture of a clock. The purpose of this activity is to expand your child's awareness of numbers in the world.

Please return this Home Link to school tomorrow.

Cut examples of numbers from scrap papers you find at home.

Glue some examples on the back of this page.

Bring examples that will not fit on this page to school.

Do not bring anything valuable!

HOME LINK
1·9

Calendars

---

1. Make a list of places at home that you find the date.

_____   _____

_____   _____

_____   _____

_____   _____

_____   _____

_____   _____

_____   _____

_____   _____

Practice

2. Write the numbers from 1 through 6.

_____ _____ _____ _____ _____ _____

Tally Marks

> **Family Note** Remind your child that the fifth tally mark crosses the other four, as follows: ⊬⊬.
>
> Counting on is an important skill that children practice whenever they count tally marks. Check that your child first counts by 5s for groups of 5 tallies and then counts by 1s. For example, ⊬⊬ ⊬⊬ ⊬⊬ /// should be counted as 5, 10, 15, 16, 17, 18. Developing this skill will take some practice.
>
> *Please return this Home Link to school tomorrow.*

1. Write 5 numbers. Make tally marks for each number.

Number	Tally Marks
18	⊬⊬ ⊬⊬ ⊬⊬ ///

Practice

2. Fill in the missing numbers on this number line.

−1   0   __   __   3   __   5   __   7   __   __

9

HOME LINK 1·11 | **Explorations**

1. Tell someone at home about your favorite mathematics Exploration. Draw something you did in your Explorations today.

Practice

Write each number.

2. ///// _____

3. ~~////~~ // _____

4. ~~////~~ ~~////~~ / _____

HOME LINK
1·12 | **Thermometers**

> **Family Note** Objects that show temperatures might be kitchen items (such as a meat thermometer) or health care items (such as a heating pad). These items do not need to show degrees Fahrenheit—they may have their own temperature gauges showing levels of heat or cold.
>
> *Please return this Home Link to school tomorrow.*

1. Look for thermometers in your home.

I found _____ thermometers in my home.

2. Do a temperature hunt. Ask someone at home to help you find other things that show temperatures.

a. Draw some of the things you find.

b. Write the name for each of your drawings. Have someone at home help you.

Practice

Write how many dots.

 **3.** _____

 **4.** _____

 **5.** _____

HOME LINK 1·13 | **Number Stories**

Family Note "Number story" is another name for what is traditionally called a "story problem" or a "word problem." *Everyday Mathematics* uses the term "number story" to emphasize the fact that the story must involve numbers.

Please return this Home Link to school tomorrow.

1. Find or draw a picture of a group of things, such as animals, people, flowers, or toys.

Have someone at home help you.

2. Tell a number story about your picture to someone at home.

3. Then glue or tape your picture to this page.

Practice

Write the number that comes before each number.

4. _____ 7   **5.** _____ 10   **6.** _____ 16

7. _____ 2   **8.** _____ 5   **9.** _____ 11

HOME LINK 1·14 | Unit 2: Family Letter

Everyday Uses of Numbers

In Unit 2, children will learn about three specific uses of numbers in real-world contexts: the use of numbers in telephone numbers, in telling time, and in counting money. Your child will learn how to interpret the various parts of a telephone number, how to tell time on the hour, and how to count collections of nickels and pennies.

When learning to tell time, your child will estimate the time on a clock that has no minute hand, only an hour hand. For example, when the hour hand is pointing exactly to the 4, we can say that the time is *about 4 o'clock.* When the hour hand is between the 4 and the 5, we can say that the time is *after 4 o'clock* or *between 4 and 5 o'clock.* By focusing on the hour hand, children will avoid the common mistake of giving the wrong hour reading (usually one hour off).

about 4 o'clock   between 4 o'clock and 5 o'clock

When learning to count money, it is preferred that your child use real coins. In Unit 2, the focus will be on counting nickels and pennies and on exchanging pennies for nickels. (In Unit 3, children will add dimes to their coin collections. Later, they will add quarters.)

Your child will also continue to develop counting skills with the help of a number grid, begin to do simple addition and subtraction problems, and continue to solve number stories.

Please keep this letter for reference as your child works through Unit 2.

Vocabulary

Important terms in Unit 2:

Numbers All Around Museum A routine that promotes number awareness and is used throughout the year as the class assembles examples of numbers used at home.

Math Boxes A collection of problems to practice skills.

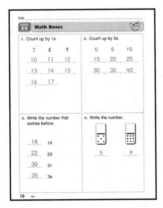

unit box A box displaying a unit, or label, in the problems at hand. For example, in problems involving counts of children in the class, the unit box would be as shown on the right.

| Unit |
| children |

analog clock A clock that shows the time by the positions of the hour and minute hands.

digital clock A clock that shows the time with numbers of hours and minutes, usually separated by a colon.

Do-Anytime Activities

To work with your child on the concepts taught in this unit and in Unit 2, try these interesting and rewarding activities:

1. Point out common uses of numbers, such as the numbers on clocks, phones, and in house addresses.

2. Help your child estimate the time using only the hour hand on an analog clock.

3. Help your child practice saying the telephone numbers of relatives and friends.

4. Count various sets of nickels and pennies together.

Have fun with these and all other mathematics activities!

Building Skills through Games

In Unit 2, your child will practice numeration and money skills by playing the following games:

Rolling for 50

Players roll a die to navigate their way on the number grid. The first player to reach FINISH wins the game!

Top-It

Each player turns over a card from the deck. Whoever has the higher number card keeps both cards. The winner is the one with more cards when the whole deck has been turned over.

Penny Plate

Starting with a plate and a specified number of pennies, one player turns the plate upside down, hiding some of the pennies. The other player counts the visible pennies and guesses how many pennies are hidden under the plate.

Penny-Nickel Exchange

Partners put 20 pennies and 10 nickels in a pile. Each player rolls a die and collects the number of pennies shown on the die. Whenever players have at least 5 pennies, they say "Exchange!" and trade their pennies for a nickel. The game ends when there are no more nickels left. The player with more nickels wins.

High Roller

Players roll two dice and keep the die with the greater number (the "high roller"). Players roll the other die again and count on from the "high roller" to get the sum of the two dice.

high roller

5, 6, 7
The sum is 7.

18

As You Help Your Child with Homework

As your child brings home assignments, you may want to go over the instructions together, clarifying them as necessary. The answers listed below will guide you through the Home Links for Unit 2.

Home Link 2·1

1–3. Check that your child's information is correct.

4. 11   **5.** 18   **6.** 20   **7.** 7

Home Link 2·2

1. Check that your child can count by 1s to the number he or she wrote.

2. Sample answer: 50, 40, 30, 20, 10, 0

3. Sample answer: I can count squares from left to right as I count by 1s. To count by 10s, I can start at the top of the last column and move down.

4. 9, 6, 5, 4, 2, 1

Home Link 2·3

1. Sample answers:

Number of Pennies in One Hand	Number of Pennies in Other Hand
5	5
4	6
2	8

2.

Number of Pennies in One Hand	Number of Pennies in the Other Hand
8	7
9	6
12	3

3. 20, 25, 30

Home Link 2·4

1. Sample answer:

Before	Number	After
8	9	10
2	3	4
0	1	2
4	5	6
5	6	7
8	9	10
10	11	12

2. 7, 8, 9, 10

Home Link 2·5

1. Sample answer:

	Tallies
Clocks	///
Watches	||||| //

Total: 10

2. Check that your child drew a picture of a clock or watch in your home.

3. 11   **4.** 15   **5.** 3

Home Link 2·6

1. Check that your child shows the hour named on his or her clock.

2. 5; 9

3.

4. ||||| //   **5.** ||||| ||||| ///   **6.** ||||| |||||

19

As You Help Your Child with Homework

As your child brings home assignments, you may want to go over the instructions together, clarifying them as necessary. The answers listed below will guide you through the Home Links for this unit.

Home Link 2·7

1. 6, 5, 8, 7, 10, 9

2. Your child's drawing should have the dominoes in order from 5 through 10.

3. 5, 6, 7, 8, 9, 10   **4.** 0, 2

5. 9, 10, 11   **6.** 16, 17, 18

Home Link 2·8

1. 5; 10; possible answer: I counted the pennies to get 5, and then counted 5 more to get 10.

2. ~H~H~H~H~H~H~H~H~H~H~H~H II

3. ~H~H~H~H~H~H~H~H~H~H~H~H~H~H

4. ~H~H~H~H~H II

5. ~H~H~H~H~H~H~H~H~H~H~H~H~H~H~H~H I

Home Link 2·9

1. Sample answer: 5, 10, 15, 20, 25, 30; 30

2. 15   **3.** 32

4. Check that your child's tally marks match his or her number.

5. 10   **6.** 21   **7.** 18   **8.** 5

Home Link 2·10

1. Ⓝ Ⓝ Ⓟ or Ⓝ Ⓟ Ⓟ Ⓟ Ⓟ Ⓟ Ⓟ; 11

2. Ⓝ Ⓝ Ⓟ; 11

3. 8   **4.** 30   **5.** 43   **6.** 17

Home Link 2·11

1. 17; 16; Sabine   **2.** 16; 20; Tony   **3.** 15

Home Link 2·12

1. 8; 11

2.

3. 21

Home Link 2·13

Check that your child has both nickels and pennies.

1. Sample answer:
2 pennies = 2¢   3 nickels = 15¢

2. Sample answer: 17¢

3. Sample answer: Toy car and pencils are circled.

 a. Sample answer: pencils   **b.** 4¢

4. Sample answers:

HOME LINK 2·1 | Telephone Numbers

> **Family Note** Work with your child to memorize important telephone numbers, including emergency daytime numbers other than your home number. Also, help your child find other examples of uses of numbers, such as:
>
> • Measurements of length, height, weight, and volume
>
> • Dates and times
>
> • Tables
>
> • Temperatures
>
> • Counts
>
> • Addresses and license plates
>
> • Costs
>
> *Please return this Home Link to school tomorrow.*

1. Write your area code and home telephone number.

(____ ____ ____) ____ ____ ____ – ____ ____ ____ ____
     (area code)   (telephone number)

2. Write an emergency number with the area code.
This number could be for a relative or a neighbor.
It might be the number for the local police department.

(____ ____ ____) ____ ____ ____ – ____ ____ ____ ____
     (area code)   (telephone number)

3. Write your first, second, and third names.

_____

Practice

Write the number that comes after each number.

4. 10 _____   **5.** 17 _____   **6.** 19 _____   **7.** 6 _____

Name _____   Date _____

Counting Up and Back

Family Note To reinforce various types of counting, listen as your child counts by 1s and 10s. Counting for someone provides good practice in this essential first-grade skill.

Please return this Home Link to school tomorrow.

1. Count for someone at home. Count up by 1s, starting with 1. I counted to _____.

2. Count back by 10s. Start with 50 or the highest number you can. I started with _____.

3. Explain to someone at home how to use the number grid to help with counts.

									0
1	2	3	4	5	6	7	8	9	10
11	12	13	14	15	16	17	18	19	20
21	22	23	24	25	26	27	28	29	30
31	32	33	34	35	36	37	38	39	40
41	42	43	44	45	46	47	48	49	50
51	52	53	54	55	56	57	58	59	60
61	62	63	64	65	66	67	68	69	70
71	72	73	74	75	76	77	78	79	80
81	82	83	84	85	86	87	88	89	90
91	92	93	94	95	96	97	98	99	100
101	102	103	104	105	106	107	108	109	110

Practice

Count back by 1s.

4. 10, _____, 8, 7, _____, _____, _____, 3, _____, _____

23

HOME LINK 2·3 | **Two-Fisted Penny Addition**

> **Family Note** By doing Two-Fisted Penny Addition, you are helping your child learn the basic addition facts. These basic facts will be useful when your child solves more difficult addition and subtraction problems mentally.
>
> *Please return this Home Link to school tomorrow.*

Do Two-Fisted Penny Addition with someone at home:

◆ On a piece of paper, draw 2 large circles.

◆ Place pennies on the table. Grab some pennies with one hand. Pick up the rest with the other hand.

◆ Place 1 pile of pennies in each circle and count them.

◆ Use the tables below to write how many pennies are in each circle.

1. Start with 10 pennies.

Number of Pennies in One Hand	Number of Pennies in the Other Hand

2. Start with 15 pennies.

Number of Pennies in One Hand	Number of Pennies in the Other Hand

---

Practice

3. Count up by 5s.

5, 10, 15, _____, _____, _____

HOME LINK 2·4 | **Numbers Before and After**

> **Family Note** When working with "before" and "after" numbers in the table below, start with small numbers—up to 15. Then, if your child is doing well, use larger numbers. You can also ask your child to suggest numbers to write in the middle column.
>
> *Please return this Home Link to school tomorrow.*

1. Ask someone to write a number in the middle column.

   ◆ Write the number that comes **before** that number.

   ◆ Write the number that comes **after** that number.

   Do this with many different numbers.

Before	Number	After
8	9	10

Practice

2. Write the numbers 7–10 below. Circle the number you wrote best.

____   ____   ____   ____

HOME LINK 2·5 Clocks and Watches

> **Family Note** In today's lesson, we observed what happens to the hour hand on an analog clock as the minute hand moves around the clock face. In the next lesson, we will practice telling time when the minute hand is pointing to 12.
>
> For the activity below, include both analog clocks (clocks that have hour hands and minute hands) and digital clocks.
>
> *Please return this Home Link to school tomorrow.*

1. Ask someone to help you find all of the clocks and watches in your home.

Record the numbers with tally marks.

	Tallies
Clocks	
Watches	

Total: _____

2. Draw a picture of the most interesting clock or watch you found. It might be interesting because of the way it looks or where it is located.

Practice

How many tally marks?

3. ⵌ ⵌ / _____   **4.** ⵌ ⵌ ⵌ _____   **5.** /// _____

29

Telling Time to the Hour

> **Family Note** We have just begun telling time to the hour. Ask your child to show times to the hour, using the paper clock.
>
> *Please return this Home Link to school tomorrow, but keep your child's paper clock for future use.*

1. Show your paper clock to someone at home.

Ask someone to name an hour for you to show.

2. Record the time.

_____ o'clock

_____ o'clock

3. Draw the hour hand.

7 o'clock

1 o'clock

Practice

Draw tally marks for each number.

4. 7 _____ **5.** 13 _____ **6.** 10 _____

Ordering Numbers

Family Note Over the next few weeks, we will "getting to know coins." In the next lesson, we will learn about pennies.

Your child is also learning how to order and compare numbers. Dominoes are a perfect tool for practicing this skill. If you have dominoes, you may want to play games with your child, such as ordering dominoes by the number of dots. At first, use consecutive numbers such as 1, 2, 3, and 4.

Please return this Home Link to school tomorrow.

Look at the dominoes below.

1. Count the total number of dots on each domino.

2. Use the back of this page. Draw the dominoes in order from the least to the greatest number of dots.

3. Write the total number of dots under each domino.

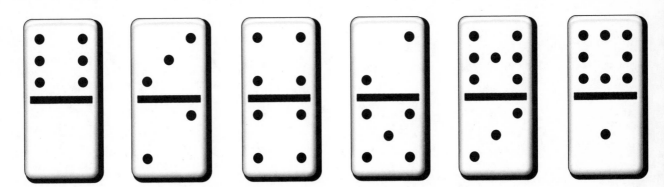

Practice

Write the numbers before and after each number.

4. _____ 1 _____   **5.** _____ 10 _____   **6.** _____ 17 _____

Nickels

> **Family Note** During the next few weeks, our class will learn about coins. For our next math lesson, your child will need to bring 5 nickels to class. Please put these nickels in a sealed envelope with your child's name on it so that they will not get lost. (Your child will also need 10 dimes and 2 quarters in the coming days.)
>
> *Please return this Home Link to school tomorrow.*

Ask someone at home for 5 nickels you can bring to school. Use one of them for this Home Link.

1. Ask someone to trade you the correct number of pennies for your nickel.

♦ How many pennies did the person give you?

_____ pennies

♦ How many pennies would you get for 2 nickels?

_____ pennies

♦ Explain to someone at home how you found your answer.

Practice

Draw tally marks for each number.

2. 27 _____

3. 35 _____

4. 17 _____

5. 41 _____

HOME LINK 2·9 Counting by 5s

Family Note Counting by 5s is a useful skill for counting combinations of coins that include nickels. A good way to practice this skill is to count tally marks.

Please return this Home Link to school tomorrow.

1. Count by 5s for someone at home.

I counted up to _____.

2. Tell someone at home how many pennies you would get for 3 nickels. _____ pennies

✔**3.** Count the tally marks below.

~~HHT~~ ~~HHT~~ ~~HHT~~ ~~HHT~~ ~~HHT~~ ~~HHT~~ //

I counted __32__ tally marks.

4. Draw some tally marks below.
Count them for someone at home.

I drew _____ tally marks.

✔ **Practice**

Write the number that is 1 less than each number.

5. 11 __10__   **6.** 22 __21__   **7.** 19 __18__   **8.** 6 __5__

HOME LINK 2·10 | **Pennies and Nickels**

Family Note First graders do not always know how to represent an amount with the fewest number of coins. That's okay. At this stage, it is important that your child understands that 5 pennies can be exchanged for 1 nickel.

In a few days, we are going to set up a "store" in our classroom. Children will take on the roles of shopkeeper and shopper. Please send some old or inexpensive items to school for our store. Thank you!

Please return this Home Link to school tomorrow.

Use ⓟ and Ⓝ to show the amount with fewer coins. Write how much the coins are worth.

Example: ⓟ ⓟ ⓟ ⓟ ⓟ ⓟ ⓟ is the same as Ⓝ ⓟ ⓟ.

This is 7 cents.

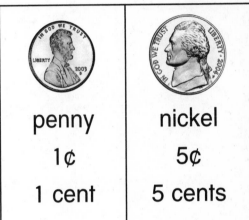

penny	nickel
1¢	5¢
1 cent	5 cents

1. ⓟ ⓟ ⓟ ⓟ ⓟ ⓟ ⓟ ⓟ ⓟ ⓟ ⓟ
is the same as

This is _____ cents.

2. Ⓝ ⓟ ⓟ ⓟ ⓟ ⓟ ⓟ is the same as

This is _____ cents.

Practice

Write the number that is 1 more than each number.

3. 7 __8__   **4.** 29 __30__   **5.** 42 __43__   **6.** 16 __17__

HOME LINK 2·11 | **Nickels and Pennies**

> **Family Note** In class, children have practiced counting combinations of pennies and nickels and then comparing amounts of money. You can use real coins to model the problems below for your child. Another way to help your child is to exchange nickels for pennies and then count the pennies.
>
> We will do a lot of work with money exchanges and with counting money. Do not expect your child to master these skills at this time.
>
> *Please return this Home Link to school tomorrow.*

1. Sabine grabbed 2 nickels and 7 pennies.

She had _____ ¢.

Tony grabbed 3 nickels and 1 penny.

He had _____ ¢.

Circle who grabbed more money: **Sabine** or **Tony**

2. Sabine grabbed 2 nickels and 6 pennies.

She had __16__ ¢.

Tony grabbed 3 nickels and 5 pennies.

He had __20__ ¢.

Circle who grabbed more money: **Sabine** or **Tony**

Practice

3. How much money? _____ ¢

HOME LINK 2·12 | **Telling Time**

1. Record the time.

_____ o'clock   _____ o'clock

2. Draw the hour hand.

9:00   6:00

Practice

3. How much money?

     _____ ¢

Counting Money

> **Family Note** This Home Link may be challenging for your child. It reviews concepts covered in this unit and applies them to new situations. Do not worry if this page is challenging—we will be working on counting money throughout the year. Encourage your child to use coins to model the problems.
>
> *Please return this Home Link to school tomorrow.*

Collect a small container of pennies and nickels.
Take a handful of the coins.

1. How many coins are in your hand? What are they worth?

_____ pennies = _____ ¢   _____ nickels = _____ ¢

2. How much are the pennies and nickels worth in all?

I counted _____ ¢ in all.

3. Circle two items that you would like to buy.

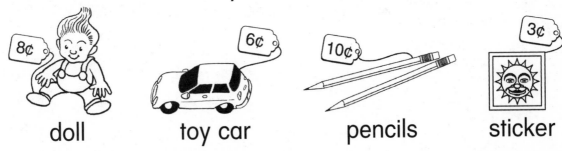

8¢   6¢   10¢   3¢

doll   toy car   pencils   sticker

a. Which item costs more? _____

b. How much more does it cost? _____ ¢ more

Practice

4. Draw 2 dominoes. Each domino should have 7 dots in all.

HOME LINK 2·14

Unit 3: Family Letter

Visual Patterns, Number Patterns, and Counting

Children will have several experiences with patterns that use objects, colors, and numbers.

Count by 10s	0, 10, 20
Count by 5s	0, 5, 10, 15, 20
Count by 2s	0, 2, 4, 6, 8, 10
Count by 3s	0, 3, 6, 9, 12

As patterns with numbers are investigated, children will look more closely at patterns found in odd and even numbers. They will observe patterns in the ending digits of counts by 2s, 3s, 5s, and 10s. Frames-and-Arrows diagrams will be introduced to help children investigate number sequences. (See explanation on next page.)

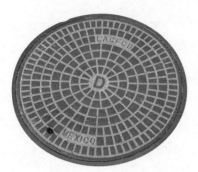

2, 4, 6, 8, 10, 12
12 is an even number.

Children also will continue to develop time-telling and money-counting skills. They will practice telling time on the hour and the half-hour. They will continue to work with real coins, so please send 10 dimes to school. (As before, please send these coins in a sealed envelope with your child's name on it.)

Finally, we will begin work on addition and subtraction. This is an important topic—it will be developed throughout the year. It is not too early for children to begin solving very simple problems.

Please keep this Family Letter for reference as your child works through Unit 3.

Vocabulary

Important terms in Unit 3:

number grid A table in which consecutive numbers are arranged in rows, usually 10 columns per row. A move from one number to the next within a *row* is a change of 1, a move of one number to the next within a *column* is a change of 10.

									0
1	2	3	4	5	6	7	8	9	10
11	12	13	14	15	16	17	18	19	20
21	22	23	24	25	26	27	28	29	30
31	32	33	34	35	36	37	38	39	40
41	42	43	44	45	46	47	48	49	50

Number grids are used to develop place-value concepts and problem-solving strategies for addition and subtraction.

pattern A repetitive order or arrangement.

even number Any counting number that ends in 0, 2, 4, 6, or 8. An even number of objects can always be grouped into pairs.

odd number Any counting number that ends in 1, 3, 5, 7, or 9. When an odd number of objects is grouped into pairs, there is always one object that cannot be paired.

Frames and Arrows Diagrams consisting of frames connected by arrows used to represent number sequences. Each frame contains one number, and each arrow represents a rule that determines which number goes in the next frame.

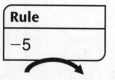

Rule
−5

35 → 30 → 25 → 20 → 15 → 10

The Family Note on Home Link 3-8, which you will receive later, provides a more detailed description of Frames and Arrows.

48

Do-Anytime Activities

To work with your child on concepts taught in this unit and in previous units, try these interesting and rewarding activities:

1. Count and pair objects found around the house and determine whether there is an odd or even number of items.

2. Using the same collection of objects, arrange them to make an ongoing pattern. Then have your child make and describe his or her own pattern.

3. Using the number grid, select a number and have your child point to the number that is 1 more or 1 less than the selected number. Or do problems like this: "Start at 28. Count back (or up) 5 spaces. On which number do you land?"

Counting back from 28

−9	−8	−7	−6	−5	−4	−3	−2	−1	0
1	2	3	4	5	6	7	8	9	10
11	12	13	14	15	16	17	18	19	20
21	22	23	24	25	26	27	28	29	30
31	32	33	34	35	36	37	38	39	40
41	42	43	44	45	46	47	48	49	50

Building Skills through Games

In this unit, your child will be practicing counting on a number line, exchanging coins, and adding by playing the following games:

Bunny Hop

Players begin at 0 on a number line marked from 0 to 20. Players take turns rolling a die and hopping the number of spaces equal to the number of dots shown on the die. The first player to hop to 20 and then back to 0 wins the game. Players must use an exact roll to land on 20 and on 0.

Coin-Dice

Players take turns rolling two dice and picking up the number of pennies equal to the number of dots shown on the dice. Whenever possible, players exchange 5 pennies for 1 nickel, 10 pennies for 1 dime, or 2 nickels for 1 dime. To pick up the last coins, the number of dots on the dice must match the number of remaining pennies.

Domino Top-It

Each player turns over a domino and calls out the sum of the dots on the domino. The player with the higher sum keeps both dominos. If there is a tie, each player chooses another domino. The player with the higher sum keeps all of the dominos. The player with more dominos at the end of the game wins.

49

As You Help Your Child with Homework

As your child brings home assignments, you may want to go over the instructions together, clarifying them as necessary. The answers listed below will guide you through the Home Links for this unit.

Home Link 3·1

2. If possible, help your child find an article of clothing with a pattern that he or she can wear to school.

3. 50, 40, 30, 20, 10

4. 25, 20, 15, 10, 5

Home Link 3·2

1. Sample answer: 4 people; even

2. Sample answer:
odd: 3, 7, 13, 19
even: 2, 6, 12, 20

3. 3 beds, odd

4. 15, 20, 25, 30, 35

5. 55, 60, 65, 70, 75

6. 95, 100, 105, 110, 115

Home Link 3·3

1. 6   **2.** 7

3. 10   **4.** 15

5. 12; 15; 16; 18   **6.** 74; 77; 78; 80

Home Link 3·4

1. Sample answer: 1648; even

2. Sample answer: odd

3. 14; ＨＨＴ ＨＨＴ ＨＨＴ ////

4. 23; ＨＨＴ ＨＨＴ ＨＨＴ ＨＨＴ ///

5. 29   **6.** 36

Home Link 3·5

1. 10, 20   **2.** 5, 10, 15, 20

3. 2, 4, 6, 8   **4.** 3, 6, 9, 12

5. All odd numbers on the number line should be circled.

6. 12   **7.** 9

Home Link 3·6

1. 7, 7   **2.** 7, 7

3. 5, 5   **4.** 16, 16

5. 6; 8; 12; 14

Home Link 3·7

7. 9   **8.** 8

Home Link 3·8

1. 7, 11, 15   **2.** 17, 14, 13

3. 15, 20, 25   **4.** 28; 30; 32; 36

Home Link 3·9

1. Add 2   **2.** Add 5   **3.** Subtract 3

4. (18)

Home Link 3·10

1. 5, 2, 10   **2.** 13, 19, 30

3. Clock should show 7:30.

4. Clock should show 3:30.

Home Link 3·11

1. 2 dimes   **2.** Ⓓ, 10

3. Ⓓ Ⓟ Ⓟ, 12

4. Ⓓ Ⓟ Ⓟ Ⓟ Ⓟ, 14

5. Ⓓ Ⓓ Ⓝ, 25   **6.** Ⓓ Ⓓ, 20

7. 30; 40; 60; 70; 80

Home Link 3·12

1. 25; 0.25   **2.** 45; 0.45

3. 23; 0.23   **4.** 37; 0.37

5. Sample answer: 2, 4, 6, 8

Home Link 3·13

2a. blue   **2b.** yellow

3. 13

Home Link 3·14

1. Sample answer: 3, 9, 15, 23

50

Name _____ Date _____

Patterns

> **Family Note** Patterns are so important in mathematics that mathematics is sometimes called the "Science of Patterns." Help your child identify patterns in your home and community.
>
> Some suggested places:
>
> - floor tiles
> - carpeting
> - window panes
> - curtains
> - wallpaper
> - fences
>
> *Please return this Home Link to school tomorrow.*

1. Find at least two patterns in your home. Draw the patterns you find on the back of this paper.

2. If you have articles of clothing (such as a shirt or a pair of socks) that have patterns, please wear them to school tomorrow!

Practice

3. Count back by 10s.

70, 60, _____, _____, _____, _____, _____

4. Count back by 5s.

35, 30, _____, _____, _____, _____, _____

HOME LINK 3·2 — Odd and Even Numbers

Family Note   As children learn about odd and even numbers, find the number of people or the number of various objects at home. Have your child tell whether these numbers are even or odd.

Please return this Home Link to school tomorrow.

1. Count the number of people in your home.

There are _____ people in my home.

Is this number **even** or **odd**? _____

2. Tell someone at home about odd and even numbers.

Write some **odd** numbers: _____, _____, _____, _____.

Write some **even** numbers: _____, _____, _____, _____.

3. Count the number of a type of object in your home.
Write the number and the type of object.

There are _____ _____ in my home.

Is this number **even** or **odd**? _____

Practice

Count up by 5s.

4. 5, 10, _____, _____, _____, _____, _____

5. 45, 50, _____, _____, _____, _____, _____

6. 85, 90, _____, _____, _____, _____, _____

HOME LINK 3·3 **Number-Line Hops**

Use the number line on the side of this page to help you answer the questions.

Example:

Start at 5. Count the hops to 11. How many hops? _6_

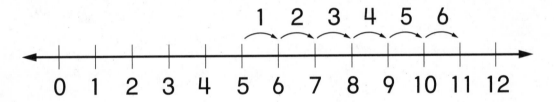

1. How many hops from 4 to 10? _____

2. How many hops from 8 to 15? _____

3. How many hops from 9 to 19? _____

4. How many hops from 1 to 16? _____

| **Practice** |

Count by 1s.

5. 11, _____, 13, 14, _____, _____, 17, _____

6. 73, _____, 75, 76, _____, _____, 79, _____

0
1
2
3
4
5
6
7
8
9
10
11
12
13
14
15
16
17
18
19
20
21
22
23
24
25

HOME LINK 3·4 | More Odd and Even Numbers

> **Family Note** We are learning to identify even and odd numbers by looking at the last digit in a number. All even numbers end in 0, 2, 4, 6, or 8. All odd numbers end in 1, 3, 5, 7, or 9. Ask your child to explain how to tell whether a number is even or odd. Give examples of odd and even numbers for your child to identify.
>
> *Please return this Home Link to school tomorrow.*

1. Write the number part of your address below.

_____

Is this number **odd** or **even**? _____

Tell someone how you know.

2. Are the addresses across the street odd or even?

_____

3. Write an even number less than 50. Show it with tally marks.

_____

4. Write an odd number less than 50. Show it with tally marks.

_____

Practice

Tell how many.

5. ~~HHH~~ ~~HHH~~ ~~HHH~~ ~~HHH~~ ~~HHH~~ IIII _____

6. ~~HHH~~ ~~HHH~~ ~~HHH~~ ~~HHH~~ ~~HHH~~ ~~HHH~~ ~~HHH~~ I _____

HOME LINK 3·5

Number Lines and Counting Patterns

Family Note	Listen as your child tells you about number lines and counts. Be sure he or she records the numbers counted. Provide several objects, such as pennies, for your child to use to count by 10s, 5s, 2s, and 3s.
	Please return this Home Link to school tomorrow.

Tell someone at home what you know about number lines and counting patterns.

Count by 10s, 5s, 2s, and 3s. Begin at 0 each time.

1. Count by 10s. 0, _____, _____

2. Count by 5s. 0, _____, _____, _____, _____

3. Count by 2s. 0, _____, _____, _____, _____

4. Count by 3s. 0, _____, _____, _____, _____

5. Circle all of the odd numbers on the number line.

Practice

What time is it?

6. _____ o'clock

7. _____ o'clock

0
1
2
3
4
5
6
7
8
9
10
11
12
13
14
15
16
17
18
19
20

HOME LINK 3·6

More Number-Line Hops

Family Note
We are working with number models like 3 + 2 = 5 and 8 − 5 = 3. We are solving them by counting up and back on the number line. Ask your child to show you how to do this. You may wish to make up number stories that use these numbers to assist your child.

For example, for 4 + 3 = _____, use the following story: "You have 4 pennies. I give you 3 more pennies. How many pennies do you have now?" Your child can use real pennies to find the answer.

Please return this Home Link to school tomorrow.

Use the number line to help you solve these problems.

1. Start at 4. Count up 3 hops. Where do you end up?

_____ $4 + 3 =$ _____

2. Start at 12. Count back 5 hops. Where do you end up?

_____ $12 - 5 =$ _____

3. Start at 11. Count back 6 hops. Where do you end up?

_____ $11 - 6 =$ _____

4. Start at 14. Count up 2 hops. Where do you end up?

_____ $14 + 2 =$ _____

Practice

Count up by 2s.

5. 2, 4, _____, _____, 10, _____, _____

0
1
2
3
4
5
6
7
8
9
10
11
12
13
14
15
16
17
18
19
20
21
22
23
24
25

LESSON 3·7 | # Telling Time to the Half-Hour

Family Note We have begun telling time to the nearest half-hour. Help your child complete these pages. Tell your child at which times, on the hour or half-hour, he or she wakes up and goes to bed on school days. Have your child tell the time at home when it is close to the hour or half-hour.

Please return these Home Link pages to school tomorrow.

Record the time.

1.

_____ o'clock

2.

half-past _____ o'clock

3.

half-past _____ o'clock

4.

half-past _____ o'clock

HOME LINK 3·7 | **Telling Time to the Half-Hour** *cont.*

Draw the hour hand and the minute hand to show the time.

5. This is about the time I wake up in the morning on a school day.

6. This is about the time I go to bed at night before a school day.

Practice

How many dots?

7. 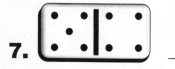 _____

8. _____

Frames-and-Arrows Diagrams

Family Note Your child is bringing home an activity you may not be familiar with. It is called "Frames and Arrows."

Frames-and-Arrows diagrams are used with sequences of numbers that follow one after the other according to a special rule. Frames-and-Arrows diagrams are made up of shapes, called **frames,** that are connected by **arrows.** Each frame contains one of the numbers in the sequence. Each arrow stands for the rule that tells how to find which number goes in the next frame. Here is an example of a Frames-and-Arrows diagram:

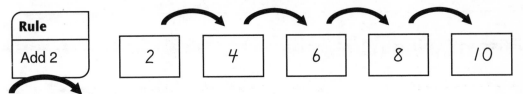

The arrow rule is "Add 2" or "Count by 2s."

In the two examples below, some of the information is left out. To solve the problem, you have to find the missing information.

Example 1: Fill in the empty frames according to the arrow rule.

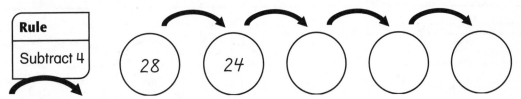

Solution: Write 20, 16, and 12 in the frames that follow 24.

Example 2: Write the arrow rule in the empty box.

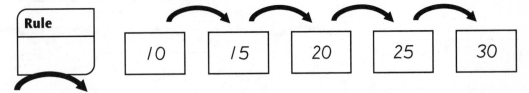

Solution: The arrow rule is "Add 5" or "Count by 5s."

Your child has been solving problems like the one in Example 1—problems in which the arrow rule is given and some of the numbers in the frames are missing. In the next lesson, we will start doing problems like Example 2, in which the numbers in the frames are given and the arrow rule is missing.

HOME LINK 3·8 **Frames-and-Arrows Diagrams** *cont.*

> **Family Note** Ask your child to tell you about Frames and Arrows. Play Frames and Arrows with him or her: One player makes up a Frames-and-Arrows problem; the other player solves it.
>
> *Please return this page of the Home Link to school tomorrow. Save page 73 for future reference.*
>
> SPECIAL NOTE: We will continue to practice counting real money in class. Please send 10 dimes to school for your child's tool-kit coin collection. We will use the dimes in 2 or 3 days. As usual, please send the coins in a securely fastened envelope with your child's name printed on the outside. Thank you!

Find the missing numbers.

1.

Rule
Add 2

| 5 | | 9 | | 13 | |

2.

Rule
Count back by 1s

| 18 | | 16 | 15 | | |

3.

Rule
5 more

| 5 | 10 | | | | 30 |

Practice

4. Count up by 2s.

24, 26, _____, _____, _____, 34, _____

Find the Rule

Family Note Today we worked with Frames-and-Arrows diagrams in which the rule was missing. You may want to refer back to the Family Note for Lesson 3-8 and review the Frames-and-Arrows routine.

Please return this Home Link to school tomorrow.

Show someone at home how to find the rules.
Then write each rule.

1.

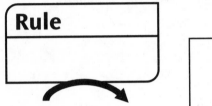

Rule

3 5 7 9 11

2.

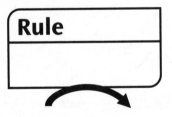

Rule

5 10 15 20 25

3.

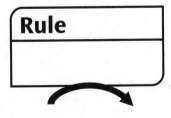

Rule

18 15 12 9 6

| **Practice** |

4. Circle the winning card in *Top-It.*

18   12

HOME LINK 3·10

Dimes

Family Note Note that Ⓟ means "penny," Ⓝ means "nickel," and Ⓓ means "dime."

IMPORTANT: Please send 10 dimes with your child to class tomorrow.

Please return this Home Link to school tomorrow.

1. How many?

_____ Ⓟ = 1 Ⓝ

_____ Ⓝ = 1 Ⓓ

_____ Ⓟ = 1 Ⓓ

2. How much money?

Ⓝ Ⓝ Ⓟ Ⓟ Ⓟ = _____ ¢

Ⓟ Ⓝ Ⓝ Ⓝ Ⓟ Ⓟ Ⓟ = _____ ¢

Ⓝ Ⓝ Ⓝ Ⓝ Ⓝ Ⓝ = _____ ¢

Practice

Draw the hour hand and the minute hand on each clock.

3.

half-past 7 o'clock

4.

half-past 3 o'clock

69

Coin Exchanges

Family Note First graders do not always know how to represent a given amount of money with the fewest number of coins. At this stage, it is important that your child understands that 5 pennies can be exchanged for 1 nickel and that 10 pennies can be exchanged for 2 nickels or 1 dime.

Please return this Home Link to school tomorrow.

1 cent 1¢	5 cents 5¢	10 cents 10¢
Ⓟ	Ⓝ	Ⓓ

1. Tell or show someone at home how many dimes you get for 4 nickels.

Show each amount below using the fewest coins. Use Ⓟ, Ⓝ, and Ⓓ.

(*Hint:* Exchange pennies for nickels and nickels for dimes.) Then write how much the coins are worth.

Example: Ⓟ Ⓟ Ⓟ Ⓟ Ⓟ Ⓟ is the same as Ⓝ Ⓟ.

This is 6 cents.

2. Ⓟ Ⓟ Ⓟ Ⓟ Ⓟ Ⓟ Ⓟ Ⓟ Ⓟ Ⓟ is the same as _____.

This is _____ cents.

Coin Exchanges *continued*

3. Ⓟ Ⓟ Ⓟ Ⓟ Ⓟ Ⓟ Ⓟ Ⓟ Ⓟ Ⓟ Ⓟ Ⓟ is the same

as  _____.

This is __12__ cents.

4. Ⓝ Ⓟ Ⓟ Ⓟ Ⓟ Ⓟ Ⓟ Ⓟ Ⓟ Ⓟ is the same

as  _____.

This is __14__ cents.

5. Ⓝ Ⓝ Ⓝ Ⓟ Ⓟ Ⓟ Ⓟ Ⓟ Ⓟ Ⓟ Ⓟ Ⓟ is the

same as _____.

This is __25__ cents.

6. Ⓝ Ⓟ Ⓝ Ⓟ Ⓝ Ⓟ Ⓟ Ⓟ is the same as _____.

This is __20__ cents.

Practice

Fill in the missing numbers.

7. 10, 20, __30__, __40__, 50, __60__, __70__, __80__

HOME LINK 3·12 Counting Coins

161 ✓

Family Note — We have counted combinations of pennies, nickels, and dimes. We are also using dollars-and-cents notation, for example, $0.52. Help your child with the problems on this page. If your child has trouble recording the amounts in dollars-and-cents notation, don't worry—this is a skill we will continue to work on throughout the year.

Please return this Home Link to school tomorrow.

1 cent 1¢ $0.01 (P)	5 cents 5¢ $0.05 (N)	10 cents 10¢ $0.10 (D)

How much money? Write each answer in cents and in dollars-and-cents.

1. (D)(N)(N)(N) _____ ¢ or $_____

2. (D)(D)(N)(N)(N)(N)(N) _____ ¢ or $_____

3. (D)(N)(N)(P)(P)(P) _____ ¢ or $_____

4. (D)(D)(N)(N)(N)(P)(P) _____ ¢ or $_____

Practice

5. Write 4 even numbers.

_____ _____ _____ _____

HOME LINK 3·13 | **Favorite Colors**

> **Family Note** Today we made a line plot for our class like the one below. At this time, your child should begin to see that the tallest column shows the color chosen by the greatest number of people and the shortest column shows the color chosen by the fewest number of people.
>
> *Please return this Home Link to school tomorrow.*

1. Tell someone at home about the line plot your class made today.

2. In Keisha's class, children made a line plot for their favorite colors.

a. What was their **favorite** color?

_____

b. What was their **least favorite** color?

_____

Favorite Colors

			B
			B
			B
R	G		B
R	G		B
R	G		B
R	G	Y	B
R	G	Y	B
R	G	Y	B
red	**green**	**yellow**	**blue**

Practice

3. How much money?

Ⓓ Ⓟ Ⓟ Ⓟ _____ ¢

HOME LINK 3·14

Domino Top-It

> **Family Note** Today your child examined dot patterns on dominoes and played with dominoes. The relationship between the numbers of dots on each domino part is useful for learning basic facts.
>
> *Domino Top-It* is a great game for helping your child practice basic addition facts.

Show someone at home how to play *Domino Top-It*. Use a set of real dominoes, if you have one. Or use the paper dominoes your teacher gave you.

Directions

1. If you have real dominoes, turn them facedown on the table. If you are using paper dominoes, put them facedown in a stack.

2. Each player takes a domino and turns it over. If you are using paper dominoes, take one from the top of the stack.

3. The player with the larger total number of dots takes both dominoes. First estimate; then count.

4. In case of a tie, each player turns over another domino. The player with the larger total takes all of the dominoes that are faceup.

5. The game is over when all of the dominoes have been played. The player who has more dominoes wins.

Practice

6. Write 4 odd numbers.

_____ _____ _____ _____

Unit 4: Family Letter

Measurement and Basic Facts

Unit 4 focuses primarily on length measurement. Lesson activities will provide opportunities for children to measure with nonstandard units, such as hand spans and paces, as well as with standard units, such as feet and inches, using a ruler and a tape measure.

Baby

0   10   20   30   40   50

Children will practice basic measuring skills, such as marking off units "end to end," aligning the 0-mark of a ruler with one edge of the object being measured, and measuring objects longer than the ruler.

Since most measurements are estimates, you will notice that estimation is used to report measurements. For example, *about* 5 hand spans, *a little less than* 8 inches, *almost* 3 feet, and so on.

Children will also practice other measurement skills. Children will read thermometers that have marks at two-degree intervals, and they will tell time to the nearest quarter-hour. Children will also explore timelines to develop a sense for sequencing events with the passage of time.

In this unit, children make number scrolls by writing numbers in extended number grids. This activity not only provides practice with writing numbers, but helps children develop a sense of the patterns in our place-value system.

-9	-8	-7	-6	-5	-4	-3	-2	-1	0
1	2	3	4	5	6	7	8	9	10
11	12	13	14	15	16	17	18	19	20
21	22	23	24	25	26	27	28	29	30
31	32	33	34	35	36	37	38	39	40
41	42	43	44	45	46	47	48	49	50
51	52	53	54	55	56	57	58	59	60
61	62	63	64	65	66	67	68	69	70
71	72	73	74	75	76	77	78	79	80
81	82	83	84	85	86	87	88	89	90
91	92	93	94	95	96	97	98	99	100

In the last two lessons, children will work toward developing addition "fact power." Knowing the basic facts is as important to mathematics as knowing words by sight is to reading. This beginning work uses dominoes as models.

$$1 + 6 = 7$$

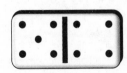

$$2 + 3 = 5$$

$$5 + 4 = 9$$

Please keep this Family Letter for reference as your child works through Unit 4.

Vocabulary

Important terms in Unit 4:

inch and foot Units of length in the U.S. customary system.

standard unit A unit of measure that has been defined by a recognized authority, such as a government or a standards organization. For example, inches and feet are standard units.

timeline A number line showing when events took place.

number scroll A series of number grids taped together.

fact power A term for the ability to automatically recall arithmetic facts without having to figure them out.

addition facts The 100 possible sums of two 1-digit numbers—from 0 + 0 through 9 + 9.

Do-Anytime Activities

To work with your child on the concepts taught in this unit and in previous units, try these interesting and rewarding activities:

1. Use a standard measuring tool to measure the length of objects in your home to the nearest inch.

2. Practice counting by 2s using a thermometer.

3. Tell the time (on the hour, the half-hour, or the quarter-hour) and have your child draw a picture of a clock to represent each time.

4. Have your child tell you the time as minutes after the hour. *For example:* "It is about six-fifteen" or "It is about fifteen minutes after six."

5. Have your child tell you a number story for a given number sentence, such as 3 + 5 = 8. *For example:* "I had 3 dogs. Then I got 5 more dogs. Now I have 8 dogs!"

I had 3 dogs. Then I got 5 more dogs. Now I have 8 dogs!

Building Skills through Games

In Unit 4, your child will play the following games:

Dime-Nickel-Penny Grab Players mix 10 dimes, 8 nickels, and 20 pennies together. One player grabs a handful of coins. The other player takes the coins that are left. Each player calculates the value of his or her coins. The player with the larger total wins the round.

High Roller Players roll two dice and keep the die with the greater number (the "high roller"). Players roll the other die again and count on from the "high roller" to get the sum of the two dice.

Shaker Addition Top-It Each player rolls two dice and calls out the sum of the dots. The player with the higher sum takes a penny. If there is a tie between players, each of these players takes a penny. The player with more pennies at the end of the game wins.

As You Help Your Child with Homework

As your child brings home assignments, you may want to go over the instructions together, clarifying them as necessary. The answers listed below will guide you through the Home Links for Unit 4.

Home Link 4·1

1. 22, 24, 26   **2.** 72, 74, 76   **3.** 52, 54, 56

4. 102, 104, 106   **5.** 70°F   **6.** 60°F

7. 80°F   **8.** 58°F

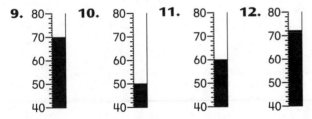

13. ~HHT~ ~HHT~ ~HHT~ /   **14.** ~HHT~ ~HHT~ ~HHT~ ~HHT~ ~HHT~

Home Link 4·2

1.–2. Your child should give a reasonable answer for how many hand spans across and long his or her bed measures.

3. 11   **4.** 10

Home Link 4·3

1.–2. Your child should make marks on the foot long foot that are about the length of each family member's foot.

3. Sample answer: It is not a good idea for people to use their own feet to measure things because everybody's feet are not the same length.

4.–5. Your child should clearly write the numbers 8 and 9.

Home Link 4·4

1. 4 **2.** 3 **3.** 12 **4.** 21

5. ~~HHT~~ ~~HHT~~ ~~HHT~~ IIII **6.** ~~HHT~~ ~~HHT~~ ~~HHT~~ ~~HHT~~ ~~HHT~~

Home Link 4·5

1.–2. Your child should measure 2 objects to the nearest inch.

3. 21¢

Home Link 4·6

1.–3. Your child should name and draw 3 measuring tools in your home such as a measuring cup, scale, or ruler.

4. odd **5.** even

Home Link 4·7

1. 10 **2.** 12

3. 11 **4.** 9

5. 9 **6.** 9

7. 25, 30, 40, 50, 55, 65, 70

8. 90, 100, 110, 120, 130

Home Link 4·8

1. 1 **2.** 4 **3.** 8 **4.** 7 **5.** 6; 5; 9

Home Link 4·9

1. Your child should draw a picture of something that happens in your family for each day of the week.

2. 16 **3.** 19 **4.** 31 **5.** 40

Home Link 4·10

1. Sample answer: I counted by 1s and wrote one number in each square as I moved from left to right on the number grid. I taped number grids together to create a scroll.

2. Sample answer: window shades or papyrus scrolls.

3.

									100
101	102	103	104	105	106	107	108	109	110
111	112	113	114	115	116	117	118	119	120
121	122	123	124	125	126	127	128	129	130

4. 23, 0.23; **5.** 41, 0.41

Home Link 4·11

1. 6 **2.** 7 **3.** 7 **4.** 5 **5.** 5 **6.** 6

7. 7 **8.** 8 **9.** 7

10.

Home Link 4·12

$$2 + 4 = 6; \begin{array}{r} 4 \\ +4 \\ \hline 8 \end{array} ; \begin{array}{r} 2 \\ +7 \\ \hline 9 \end{array} ; 8 = 5 + 3; 7 = 4 + 3;$$

$$\begin{array}{r} 1 \\ +6 \\ \hline 7 \end{array} ; \begin{array}{r} 6 \\ +3 \\ \hline 9 \end{array} ; 8 + 2 = 10; 10 = 6 + 4$$

6, 8, 12

HOME LINK 4·1

Reading Thermometers

Fill in the frames.

1.

Rule
Count by 2s

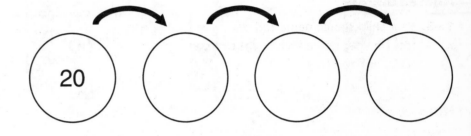

2.

Rule
Count by 2s

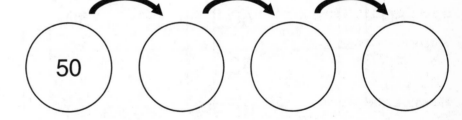

3.

Rule
Count by 2s

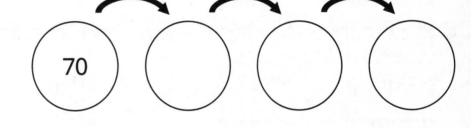

4.

Rule
Count by 2s

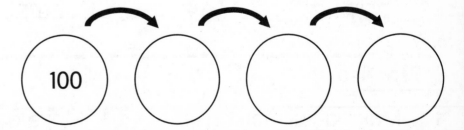

Reading Thermometers *continued*

Write the temperature shown by each thermometer.
Write °F with the temperature.

5.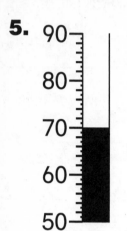

6.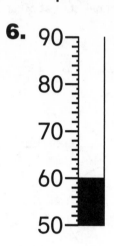

7.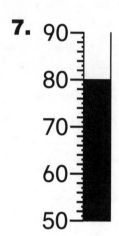

8.

_____   _____   _____   _____

Color the thermometer to show each temperature.

9.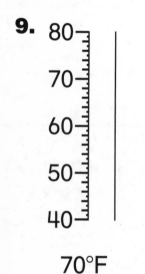

10.

11.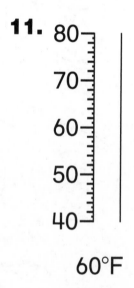

12.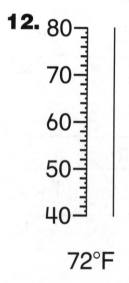

70°F   50°F   60°F   72°F

Practice

13. Make a tally for 16.   **14.** Make a tally for 25.

_____   _____

HOME LINK 4·2 **Measuring with Hand Spans**

Family Note In today's lesson, we measured objects using nonstandard units such as digits (finger widths), hands, fathoms (arm spans), and hand spans.

  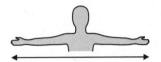

One of our discoveries is that we get different measurements for the same object if different people do the measuring using these units. We will continue this investigation, eventually realizing that standard units, such as feet and inches, provide us with more reliable measurements. Help your child measure his or her bed using hand spans. See drawing below.

Please return this Home Link to school tomorrow.

Measure your bed with your hand span.

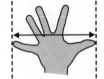

1. How many hand spans across is it?

About _____ hand spans

2. How many hand spans long is it?

About _____ hand spans

Practice

3. Start at 7. Move 4 hops. Where do you end up? _____

4. Start at 5. Move 5 hops. Where do you end up? _____

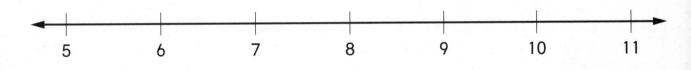

The Foot-Long Foot

> **Family Note** To help us investigate the measuring unit "feet," please help your child mark each family member's foot on page 97, using different-colored crayons.
>
> *Please return this Home Link to school tomorrow.*

Compare the foot-long foot to the feet of members of your family.

Here is what you do:

1. Mark the length of each person's foot onto the foot-long foot. Use a different-colored crayon for each person's foot.

2. Label each mark with the person's name.

3. Talk about why it is not a good idea for people to use their own feet for measuring things.

Practice

Practice writing the numbers 8 and 9.

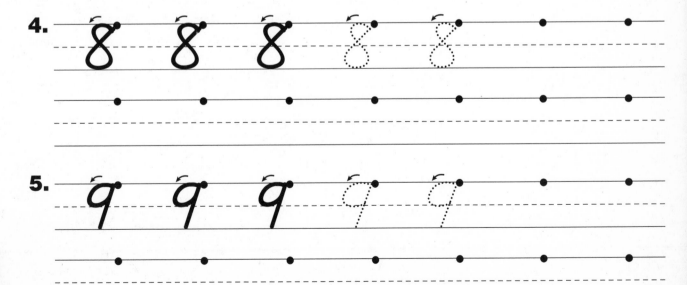

4.

5.

HOME LINK 4·4 # Measuring with Inches

Family Note We are learning how to measure to the nearest inch. Do not expect your child to be proficient with this skill at this time. We will be practicing throughout the year. To help your child, make sure that he or she lines up one end of the object with the 0-mark on the ruler. Help your child find the closest inch mark at the other end of the object.

Please return this Home Link to school tomorrow.

Use your 12-inch ruler to measure the objects below. Record your measurements.

1.

About _____ inches long

2. CRAYON

About _____ inches long

Practice

How many tally marks?

3. ⵑⵑⵑ ⵑⵑⵑ || _____

4. ⵑⵑⵑ ⵑⵑⵑ ⵑⵑⵑ ⵑⵑⵑ | _____

5. Draw tally marks to show 19.

_____

6. Draw tally marks to show 25.

_____

Measuring with a Ruler

Family Note	This activity is the same as the activity on the previous Home Link, except that this time your child will choose objects to measure.
	Have your child measure objects to the nearest inch. Make sure your child lines up one end of the object being measured with the 0-mark on the ruler.
	Please return this Home Link to school tomorrow.

Use your 12-inch ruler to measure 2 small objects to the nearest inch. Draw a picture of each object. Record your measurements.

1.

About _____ inches long

2.

About _____ inches long

Practice

3. How much money?

(N)(N)(N)(P)(P)(P)(P)(P)(P) _____ ¢

Measuring Tools

Family Note We have been working with linear measures, using rulers and tape measures. The length of an object is an example of a linear measure.

Help your child find other kinds of measuring tools in your home, such as scales that measure weight, measuring cups that measure capacity, and so on.

Please return this Home Link to school tomorrow.

Name and draw 3 measuring tools in your home.

Example:

1. _____

2. _____

3. _____

Practice

Odd or even?

4.

5.

_____   _____

HOME LINK 4·7 | **Domino Dots**

Draw the missing dots on each domino.
Write the total number of dots.

1. _____10_____

8   2

2. _____12_____

6   6

3. _____11_____

5   6

4. _____9_____

4   5

5. _____9_____

7   2

6. _____9_____

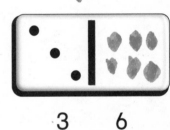

3   6

Practice

7. Count up by 5s.

10, 15, 20, __35__, __30__, 35, __40__, 45, __50__,

__55__, 60, __65__, __70__, 75

8. Count up by 10s.

60, 70, 80, __90__, __100__, __110__, __120__, __130__

95

HOME LINK 4·8 | # Telling Time

> **Family Note** We have been learning to tell time on the hour and the half-hour. Today we began to learn how to tell time on the quarter-hour.
>
> *Please return this Home Link to school tomorrow.*

Record the time.

1.

_____ o'clock

2.

half-past

_____ o'clock

3.

half-past

_____ o'clock

4.

_____ o'clock

Practice

5. Make sums of 10 pennies

Left Hand	Right Hand
3	7
4	
	5
1	

HOME LINK 4·9 My Timeline

1. Draw pictures of important things that happen in your family each day of the week.

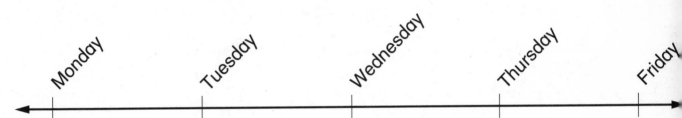

Monday   Tuesday   Wednesday   Thursday   Friday

Practice

Write the number that is one less.

2. _____ 17   **3.** _____ 20   **4.** _____ 32   **5.** _____ 41

HOME LINK 4·10 Number Grids

Family Note Ordering numbers on a grid is important in identifying number patterns and developing number power. You and your child may want to talk about patterns in the number grid shown below.

Please return this Home Link to school tomorrow.

1. Tell your family how you filled in number grids and made scrolls.

2. Ask if your family knows about any other kinds of scrolls.

3. Show someone how you can fill in the bottom 3 rows of this number grid.

									100
101									
				115					
									130

Practice

How much money? Write each answer in cents and dollars-and-cents.

 4. ⒹⓃⓃⓅⓅⓅ _23_ ¢ or $_0.23_

 5. ⒹⒹⒹⓃⓃⓅ _41_ ¢ or $_0.41_

101

Domino Sums

Find the sums.

1.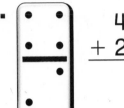
$$\begin{array}{r} 4 \\ +\ 2 \\ \hline \end{array}$$

2.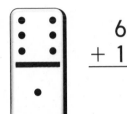
$$\begin{array}{r} 6 \\ +\ 1 \\ \hline \end{array}$$

3.
$$\begin{array}{r} 7 \\ +\ 0 \\ \hline \end{array}$$

4.

1 + 4 = ____

5.

____ = 2 + 3

6.

3 + 3 = ____

7.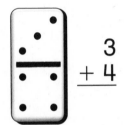
$$\begin{array}{r} 3 \\ +\ 4 \\ \hline \end{array}$$

8.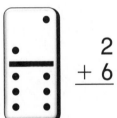
$$\begin{array}{r} 2 \\ +\ 6 \\ \hline \end{array}$$

9.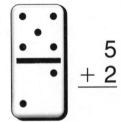
$$\begin{array}{r} 5 \\ +\ 2 \\ \hline \end{array}$$

Practice

10. Draw the next 3 shapes.

     ____ ____ ____

HOME LINK 4·12 | Color-by-Number

Family Note We are finding sums for addition facts, using +0, +1, +2 (such as 3 + 0, 5 + 1, and 8 + 2), and doubles facts (such as 2 + 2 and 4 + 4).

Please return this Home Link to school tomorrow.

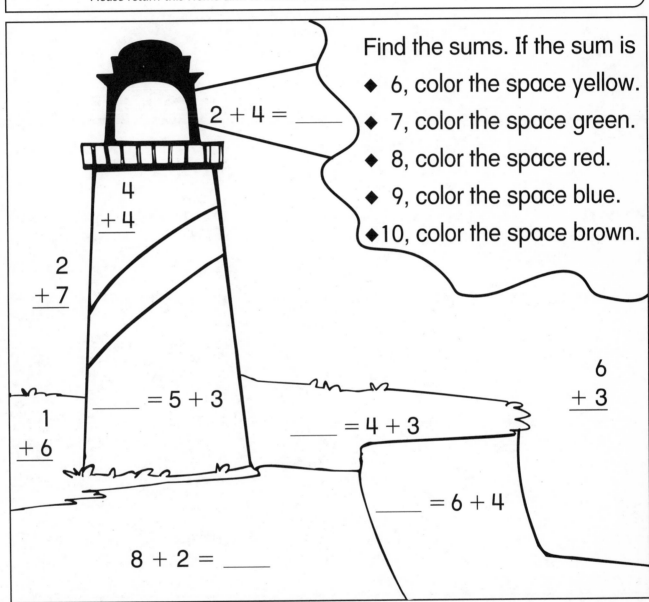

$2 + 4 =$ ____

Find the sums. If the sum is
- ◆ 6, color the space yellow.
- ◆ 7, color the space green.
- ◆ 8, color the space red.
- ◆ 9, color the space blue.
- ◆10, color the space brown.

$\begin{array}{r} 4 \\ +\,4 \\ \hline \end{array}$

$\begin{array}{r} 2 \\ +\,7 \\ \hline \end{array}$

$\begin{array}{r} 6 \\ +\,3 \\ \hline \end{array}$

____ $= 5 + 3$

____ $= 4 + 3$

$\begin{array}{r} 1 \\ +\,6 \\ \hline \end{array}$

____ $= 6 + 4$

$8 + 2 =$ ____

Practice

Write the missing numbers.

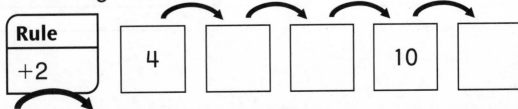

Rule					
+2	4			10	

HOME LINK 4·13 | # Unit 5: Family Letter

Place Value, Number Stories, and Basic Facts

As their work in mathematics progresses, children are beginning to use larger numbers. In Unit 5, children will begin to explore the system we use for writing large numbers by focusing on the idea of *place value*. For example, in the number 72, 7 is in the tens place, so there are "7 tens," and 2 is in the ones place, so there are "2 ones." Children will use base-10 blocks to represent numbers and to find the sums of two numbers. They will also use place value to determine "greater than" and "less than" relationships.

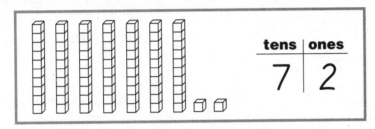

Later in this unit, children will continue to work with addition facts. Shortcuts for learning facts will be introduced. One shortcut is the *turn-around* rule, which states that the order in which numbers are added does not change the sum. For example, 4 + 3 and 3 + 4 both equal 7. Your child will also learn the meaning of adding 0 and 1 to any number. Knowing these shortcuts will make the task of learning addition facts easier.

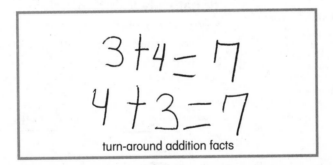

turn-around addition facts

Children will also practice place value and addition and subtraction facts by acting out number stories. They will act out these stories using concrete objects and will begin to represent the stories with *number models.* (See this unit's vocabulary list for more information on number models.)

Children have explored many number patterns in previous lessons. *"What's My Rule?"* is a routine introduced in this unit and found throughout *Everyday Mathematics* that provides practice with number patterns and number relationships. You will receive more detailed information about this routine when we begin to use it in class.

Please keep this Family Letter for reference as your child works through Unit 5.

Vocabulary

Important terms in Unit 5:

cube In *Everyday Mathematics*, a base-10 block that represents 1.

long In *Everyday Mathematics*, a base-10 block that represents 10.

flat In *Everyday Mathematics*, a base-10 block that represents 100.

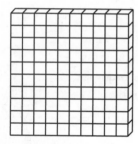

place value In our standard, base-10 system for writing numbers, each place has a value 10 times that of the place to its right and 1 tenth the value of the place to its left. For example, in the number 54, the 5 represents tens, and the 4 represents ones.

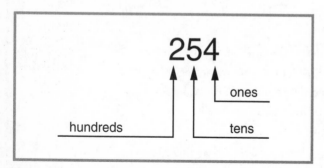

254

ones

hundreds   tens

number model A number sentence that models a number story.

For example, $7 + 3 = 10$

Unit
lions

is a number model for the number story:

Seven lions are lying in the sun. Three more lions join them. How many lions are there altogether?

turn-around addition facts A pair of addition facts in which the order of the addends is reversed. For example, $5 + 4 = 9$ and $4 + 5 = 9$ are turn-around addition facts.

doubles addition facts The sum of a 1-digit number added to itself. For example, $5 + 5 = 10$, $2 + 2 = 4$, and $6 + 6 = 12$ are all doubles addition facts. A doubles addition fact does not have a turn-around addition fact partner.

function machine An imaginary device that receives inputs and generates outputs. A number (input) is put into the machine and is transformed into a second number (output) through the application of a rule.

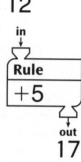

12

in

Rule
+5

out
17

"What's My Rule?" problem

A problem in which two of the three parts of a function (input, output, and rule) are known, and the third is to be found out.

Rule
+5

in	out
2	7
5	10
7	12
6	11

Do-Anytime Activities

To work with your child on the concepts taught in this unit and in previous units, try these interesting and rewarding activities:

1. Tell addition and subtraction number stories to your child. Have your child solve the problems using various household objects, and then record the answers in number models.

2. Encourage your child to make up some number stories.

Building Skills through Games

In this unit, your child will practice addition, subtraction, and place-value skills by playing the following games:

Base-10 Exchange
Players take turns putting base-10 blocks on their Tens-and-Ones Mat according to the roll of a die. Whenever possible, they exchange 10 cubes for 1 long. The first player to get 10 longs wins.

Beat the Calculator
A "Calculator" (a player who uses a calculator) and a "Brain" (a player who does not use a calculator) race to see who will be first to solve addition problems.

Difference Game
Players pick a card and collect as many pennies as the number shown on the card. Then players count each other's pennies and figure out how many more pennies one player has than the other.

Digit Game
Each partner draws two cards from a set of number cards. The player whose cards make the larger number takes all of the cards. The player with more cards at the end of the game wins.

Penny-Nickel-Dime Exchange
Partners place 20 pennies, 10 nickels, and 10 dimes into a bank. Players take turns rolling two dice, collecting the amount shown on the dice from the bank. Partners exchange pennies and nickels for dimes until all of the dimes are gone. The player who has more dimes wins.

As You Help Your Child with Homework

As your child brings assignments home, you may want to go over the instructions together, clarifying them as necessary. The answers below will guide you through the Home Links in this unit.

Home Link 5·1

1. 56 **2.** 73 **3.** 12 **4.** 60; 50

Home Link 5·2

1. 30, 40, 50, 70 **2.** 110, 100, 90, 70

3. 78, 68, 48, 38

4. Sample answer: Ⓓ Ⓓ Ⓟ Ⓟ

5. Sample answer: Ⓓ Ⓓ Ⓓ Ⓝ

Home Link 5·3

1. > **2.** < **3.** =

4. < **5.** > **6.** <

7. Answers vary. **8.** Answers vary.

Home Link 5·4

1. 32, 0.32 **2.** 36, 0.36

3. 38, 0.38

4. ~~HHT~~ ~~HHT~~ ~~HHT~~ ~~HHT~~ ~~HHT~~ ~~HHT~~, even

Home Link 5·5

1. 8 **2.** 6 **3.** 3

4. 6 **5.** 6 **6.** 9

7. 4 **8.** 8 **9.** 5

10. 4④, 3①, 1⑦, 6⑨

Home Link 5·6

1. < **2.** > **3.** =

4. < **5.** > **6.** <

7.

8.

9.

10.

Home Link 5·7

1. Bart, 4 **2.** Martha, 7 **3.** Maria, 8

4. 1①5, ⑧0, ⑤5, ①7

Home Link 5·8

1. Your child should write a number story and number model to go with his or her picture.

2. 6 **3.** 10 **4.** 6

Home Link 5·9

1. > **2.** < **3.** = **4.** =

5. 7 **6.** 9 **7.** ⑧ **8.** ⑫

Home Link 5·10

1. 6 + 3 = 9 **2.** 3 + 6 = 9

3. 5 + 4 = 9 **4.** 4 + 5 = 9

5. 24 **6.** 47

Home Link 5·11

1. Answers will vary. **2.** Answers will vary.

3. < **4.** >

5. < **6.** =

Home Link 5·12

1. Rule is +1; 20, 10, (last answer will vary)

2. Rule is −2; 10, 19, (last answer will vary)

3. Rule is +10; 35, (last answer will vary)

4. 10 **5.** 14

6. 6 **7.** 18

Home Link 5·13

1. Rule is +3, (answer will vary)

2. 16, 35, (last answer will vary)

3. Answers vary.

4. 40, 38, 36, 34, 32, 30, 28, 26, 24

Name _____   Date _____

Tens-and-Ones Riddles

> **Family Note** We have begun to work on place value using base-10 blocks. The blocks shown in the tens columns are called *longs* and the blocks shown in the ones columns are called *cubes*. It takes 10 cubes to make 1 long. On this page, your child is writing numbers shown with longs and cubes.
>
> *Please return this Home Link to school tomorrow.*

Example:

Tens	Ones

What number am I? __28__

1.

Tens	Ones

What number am I? _____

2.

Tens	Ones

What number am I? _____

3.

Tens	Ones

What number am I? _____

Practice

4. Fill in the missing numbers.

Rule
Count back by 10s

70 → _____ → _____ → 40

2) 5H

$$-\dfrac{41}{23}$$

120

Frames-and-Arrows Diagrams

Fill in the missing numbers.

1.

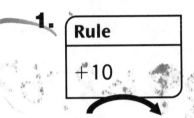

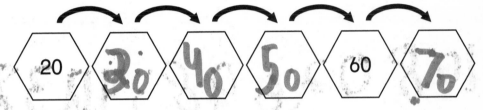

Rule: +10

20 30 40 50 60 70

2.

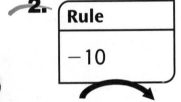

Rule: −10

120 110 ~~100~~ 90 80 70

3.

Rule: Count back by 10s

88 78 68 58 48 38

Practice

4. Show 22¢.

Use , , and .

5. Show 35¢.

Use , , and ⓟ.

108 026 9839937968767|

HOME LINK 5·3 Relation Symbols

Family Note The relation symbols < and > were introduced in this lesson. The symbol < means *is less than,* and the symbol > means *is more than.* These symbols will be used in the same way we use the symbol = for *is equal to* or *equals.* For example, instead of writing 5 *is less than* 8, we will write 5 < 8.

It takes time for children to learn the correct use of these symbols. One way to help your child identify the correct symbol is to draw two dots near the larger number and one dot near the smaller number. Then connect the dots as shown below.

$$5 < 8$$

Another way is to think of the open end of the symbol as a mouth eating the larger number.

 5 < 8

Write <, >, or =.

Example:

18 __>__ 12

< is less than
> is more than
= is the same as
= is equal to

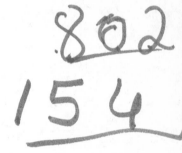
802
154

1. 11 __>__ 7   **2.** 21 __<__ 25   **3.** 37 __=__ 37

4. 29 __<__ 42   **5.** 35 __>__ 15   **6.** 48 __<__ 128

48   834

Practice

7. Write some even numbers below.

45 2 6 100 200 708

8. Write some odd numbers below.

1   3   41 227 3007 83

45291,   45 923965

every

HOME LINK 5·4 Counting Coins

Family Note Children continue finding the values of groups of coins. Before doing the problems, it may be helpful for your child to sort real coins into groups (all of the dimes together, all of the nickels together). Many children are still learning to write amounts of money using dollars-and-cents notation. We will continue to practice this skill during the year.

Please return this Home Link to school tomorrow.

Ⓟ 1 cent	Ⓝ 5 cents	Ⓓ 10 cents
$0.01	$0.05	$0.10
penny	nickel	dime

How much? Write each answer in cents and in dollars-and-cents notation.

1. Ⓓ Ⓝ Ⓝ Ⓝ Ⓝ Ⓟ Ⓟ _____ ¢ or $____ . ____

2. Ⓓ Ⓝ Ⓝ Ⓝ Ⓝ Ⓟ _____ ¢ or $____ . ____

3. Ⓓ Ⓓ Ⓝ Ⓝ Ⓝ Ⓟ Ⓟ Ⓟ _____ ¢ or $____ . ____

Practice

4. Make a tally for 30.

_____

Odd or even? _____

Quarter to 4 o'clock

3:45

HOME LINK 5·5 Domino Addition

> **Family Note** Children continue practicing basic addition facts. Notice that we are emphasizing +0, +1, and double facts like 6 + 6.
>
> *Please return this Home Link to school tomorrow.*

Add.

1.

$$\begin{array}{r} 4 \\ + 4 \\ \hline \end{array}$$

2.

$$\begin{array}{r} 0 \\ + 6 \\ \hline \end{array}$$

3.

___ = 2 + 1

4.
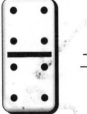

1 + 5 = ___

5.
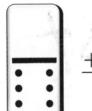

3 + 3 = ___

6.

___ = 0 + 9

Fill in the missing dots and the missing numbers.

7.

$$\begin{array}{r} 4 \\ + 0 \\ \hline \end{array}$$

8.
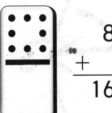
$$\begin{array}{r} 8 \\ + \\ \hline 16 \end{array}$$

9.
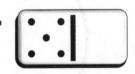

5 + ___ = 10

| **Practice** |

10. Circle the ones place.

44   31   17   69

quarter to :5

~~quarter~~ &

Atmost

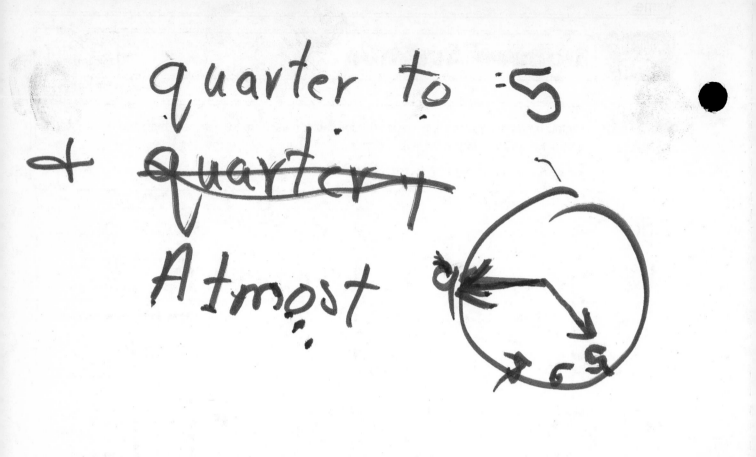

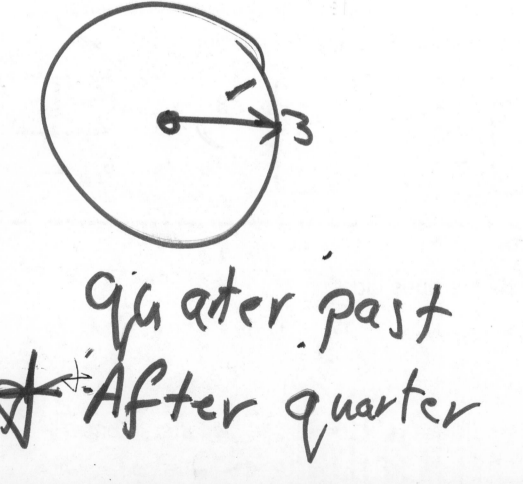

quater past

~~&~~ After quarter

HOME LINK 5·6

Relation Symbols

Family Note As children continue their work with relation symbols (<, >, =), you can help by having your child read aloud the number models on this page. Read the example as follows: 65 is less than 83.

Please return this Home Link to school tomorrow.

Write <, >, or =.

Example: 65 __<__ 83

1. 15 __<__ 17

2. 28 __>__ 19

3. 24 __=__ 24

4. 36 __<__ 63

5. 92 __>__ 72

6. 55 __<__ 128

| < is less than |
| > is more than |
| = is the same as |
| = is equal to |

Practice

Draw the hour and minute hands to show each time.

7.

eleven o'clock

8.

nine thirty

9.

half-past six

10.

quarter-to one

9 12 45

121

Name __Arvind Sehra__   Date __8/26/2010__

1. Bart   Ⓟ Ⓟ Ⓟ Ⓟ Ⓟ Ⓟ Ⓟ Ⓟ Ⓟ Ⓟ Ⓟ Ⓟ

Perry   Ⓟ Ⓟ Ⓟ Ⓟ Ⓟ Ⓟ Ⓟ Ⓟ

Who has more? __Bart__ How much more? __4__ ¢

2. Tricia   Ⓟ Ⓟ Ⓟ

Martha   Ⓟ Ⓟ Ⓟ Ⓟ Ⓟ Ⓟ Ⓟ Ⓟ Ⓟ Ⓟ

Who has more? __Martha__ How much more? __7__ ¢

3. Franklin has 17 pennies.

Maria has 25 pennies.

Who has more? __Maria__ How much more? __8__ ¢

Practice

4. Circle the tens place.

 1⑮   ⑧0   ⑤5   ①7

123

HOME LINK 5·8 Number Stories

Family Note Children have been telling and solving number stories. Have your child explain the number story that goes with the picture he or she chooses. If you like, help your child record the number story in words. The number model may show addition or subtraction, depending on how your child tells the story.

Please return this Home Link to school tomorrow.

Here is a number story Mandy made up.

I have 4 balloons.

Jamal brought 1 more.

We have 5 balloons together.

4 + 1 = 5

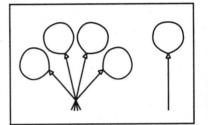

Unit
balloons

---

1. Find a picture in a magazine or draw your own picture. Use it to write a number story.

Write a number model to go with your story.

Unit

Practice

Write each sum.

2.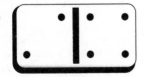

2 + 4 = _____

3.

6 + 4 = _____

4.

5 + 1 = _____

HOME LINK 5·9   # Comparing Sums

Family Note   For the next few days, children will return to basic addition facts. They will concentrate on memorizing the +0 and +1 facts (for example, 7 + 0 and 8 +1), doubles facts (for example, 3 + 3), and facts that have a sum of 10 (for example, 3 + 7 and 6 + 4). Consider spending a short time each day practicing these addition facts with your child.

Please return this Home Link to school tomorrow.

Write <, >, or =.

1. 🎲 + 🎲 __>__ 🎲 + 🎲

2. 🎲 + 🎲 __<__ 🎲 + 🎲

3. 🎲 + 🎲 __=__ 🎲 + 🎲

4. [domino] __=__ [domino]

Practice

Find the sums.

5. $4 + 3 =$ __7__   6. __9__ $= 0 + 9$

7. __8__ $= 6 + 2$ 8. __12__ $= 10 + 2$

Circle the even sums.

127

HOME LINK 5·10

Turn-Around Dice

Family Note Turn-around addition facts are pairs of facts in which the numbers being added are the same. Turn-around facts have the same sum. For example, 2 + 3 = 5 and 3 + 2 = 5 are turn-around facts. Knowing about turn-around facts cuts down on the number of facts that have to be memorized: If you know a fact, you also know its turn-around fact.

Please return this Home Link to school tomorrow.

Find the total number of dots on the dice.
Use turn-around facts to help you.

Unit
dice dots

1. +   <u> 6 </u> + <u> 3 </u> = <u> 9 </u>

2. + <u> 3 </u> + <u> 6 </u> = <u> 9 </u>

3. + + +   9

4. + + +   9

Practice

Solve the riddles.

5. 2 and 4 ⬚ = <u> 24 </u>   **6.** 4 and 7 ⬚ = <u> 47 </u>

Name _____ Date _____

Family Note Give your child several 1-digit, 2-digit, and 3-digit numbers. Ask him or her to add 0 and 1 to each number.

Include numbers with 9 in the ones place like 9, 49, 79, 129, 359, and 789. Also use 0 in the tens and ones places, like in 208 and 320.

Please return this Home Link to school tomorrow.

Record your answers in the table below.

1. Ask someone at home to say a 1-digit number; for example, 7. Add 0 to the number and give the answer. Then add 1 to the number and give the answer.

2. Have someone say a 2-digit number. Repeat with a 3-digit number.

Example: 25 + 0 = 25   25 + 1 = 26

Number Models

	Number	**+0**	**+1**
Example	25	25 + 0 = 25	25 + 1 = 26
1-digit number			
2-digit number			
3-digit number			

Practice

Write <, >, or =.

3. 19 _____ 21   **4.** 10 _____ 4   **5.** 2 _____ 11   **6.** 0 _____ 0

Family Letter

"What's My Rule?"

Today your child learned about a kind of problem you may not have seen before. We call it "What's My Rule?" Please ask your child to explain it to you. Here is a little background information you may find useful.

Imagine a machine with a funnel at the top and a tube at the bottom—we call this a *function machine.* The function machine can be programmed so that when you drop a number into the funnel at the top, the machine changes the number according to the rule and a new number comes out of the tube at the bottom.

For example, you can program the machine to add 2 to any number that is dropped into the funnel. If you put in 3, out comes 5; if you put in 6, out comes 8.

You can show this with a table:

3
in ↓

Rule

+2

out ↓
5

in	out
3 →	5
6 →	8
10 →	12

Here is another example of a function machine:

5
in ↓

Rule

+3

out ↓
8

in	out
5 →	8
6 →	9
2 →	5

In a "What's My Rule?" problem, some of the information is missing. To solve the problem, you have to find the missing information. The missing information can be the numbers that come out, the numbers that are dropped in, or the rule for programming the machine. *For example:*

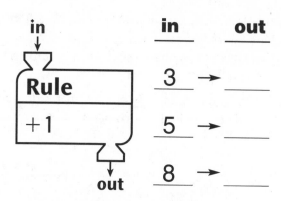

in	out
3 →	___
5 →	___
8 →	___

Missing "out" numbers

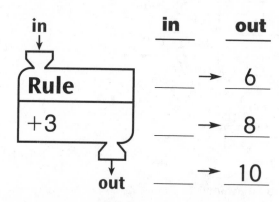

in	out
___ →	6
___ →	8
___ →	10

Missing "in" numbers

in	out
6 →	4
10 →	8
12 →	10

Missing rule

HOME LINK 5·12 "What's My Rule?"

Fill in the missing rule and numbers.

1.

in ↓

Rule
+1

out ↓

in		out
6	→	7
14	→	15
26	→	27
19	→	20
9	→	10

Your turn: 78 → 79

2.

in ↓

Rule
−2

out ↓

in		out
10	→	8
22	→	20
25	→	23
12	→	10
21	→	19

Your turn: 62 → 59

3.

in ↓

Rule
+10

out ↓

in		out
36	→	46
19	→	29
62	→	72
25	→	35

Your turn: 85 → 95

Practice

Add.

4. 5 + 5 = 10

5. 7 + 7 = 14

6. 3 + 3 = 6

7. 9 + 9 = 19

More "What's My Rule?"

> **Family Note**
> Children continue to explore number patterns. Each problem on this page represents a different kind of problem.
>
> In the first problem, your child tries to find the rule. In the second problem, the rule is given. The second problem calls for applying the rule to find the "out" numbers.
>
> Encourage your child to describe how he or she solved each problem.
>
> *Please return this Home Link to school tomorrow.*

1. Find the rule.

in ↓ → **Rule**
out

in	out
5	8
10	13
18	21

Your turn: _____

2. What comes out?

in ↓ → **Rule** −10
out

in	out
13	3
26	
45	

Your turn: _____

3. Make your own.

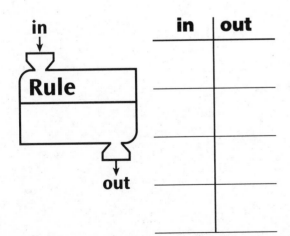

in ↓ → **Rule**
out

in	out

> **Practice**

4. Count back by 2s.

46, 44, 42,

____, ____, ____,

____, ____, ____,

____, ____, ____

Unit 6: Family Letter

Developing Fact Power

Knowing the basic facts is as important to mathematics as knowing words by sight is to reading. Your child should begin to master many addition and subtraction facts by the end of the year.

Learning the facts takes practice. It is not necessary to practice for a long time, but it is important to practice often. One good way to practice is to play the games described on the third page of this letter.

Later in this unit, children will extend their time-telling skills by learning to tell time to the nearest 5 minutes and by representing the time in digital notation, as it appears on a digital clock.

Math Tools

Your child will be using *Fact Triangles* to practice and review addition and subtraction facts. Fact Triangles are a "new and improved" version of flash cards; the addition and subtraction facts shown are made from the same three numbers, and this helps your child understand the relationships among those facts. The *Family Note* on Home Link 6-4, which you will receive later, provides a more detailed description of Fact Triangles.

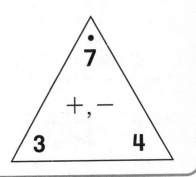

Please keep this letter for reference as your child works through Unit 6.

Vocabulary

Important terms in Unit 6:

fact family A set of related facts linking two inverse operations, such as addition and subtraction. For example:

$$3 + 4 = 7$$
$$4 + 3 = 7$$
$$7 - 3 = 4$$
$$7 - 4 = 3$$

function machine An imaginary device that receives inputs and generates outputs. The machine usually pairs an input number with an output number by applying a rule such as " +5."

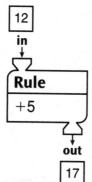

name-collection box A diagram that is used for collecting equivalent names for a number.

9	⊬⊬⊬ ////
	7 + 2 3 + 4 + 2
	9 − 0 10 − 1
	nine ••• ••• •••

digital clock A clock that shows the time with numbers of hours and minutes, usually separated by a colon.

range The difference between the maximum and minimum in a set of data. For example, in the set below, the range is $36 - 28 = 8$.

middle value The number in the middle when the data are listed from smallest to largest. For example, in the data set below, 32 is the middle value:

28 28 31 **32** 33 35 36

"What's My Rule?" problem A problem in which two of the three parts of a function (input, output, and rule) are known, and the third is to be found out.

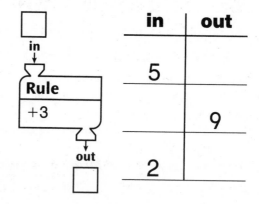

Do-Anytime Activities

To work with your child on the concepts taught in this unit and previous units, try these interesting and rewarding activities:

1. Using the Fact Triangles, cover the sum for addition practice. Cover one of the other numbers for subtraction practice. Make this brief and fun.

2. Have your child tell you a number story that fits a given number model, such as $3 + 5 = 8$.

3. Go to your local library and check out the book **12 Ways to Get to 11** by Eve Merriam, an entertaining book that presents addition facts.

4. Fill in name-collection boxes. Begin with a number, such as 20, and find at least five equivalent names.

Building Skills through Games

In Unit 6, your child will practice addition, subtraction, and money skills by playing the following games.

Addition Top-It

Players turn over two cards and call out the sum. The player with the higher sum keeps all of the cards. The player with more cards at the end of the game wins.

Fact Power Game

Players take turns naming sums of addition facts on a game board. The player who correctly answers the greatest number of addition facts wins the game.

Coin Exchange

Players put 20 pennies, 10 nickels, and 2 quarters in a pile. At each turn, a player rolls 2 dice and collects the amount of money equal to the number of dots on the dice. Players make exchanges whenever possible. The game ends when there are no more quarters. The player who has the greatest amount of money wins.

As You Help Your Child with Homework

As your child brings assignments home, you may want to go over the instructions together, clarifying them as necessary. The answers listed below will guide you through the Home Links in this unit.

Home Link 6·1

1. $\begin{array}{r} 5 \\ +9 \\ \hline 14 \end{array}$ (yellow); $6 + 6 = 12$ (blue);
    $7 + 7 = 14$ (yellow)

    $\begin{array}{r} 8 \\ +7 \\ \hline 15 \end{array}$ (red); $12 = 3 + 9$ (blue); $\begin{array}{r} 3 \\ +7 \\ \hline 10 \end{array}$ (green); $\begin{array}{r} 5 \\ +7 \\ \hline 12 \end{array}$ (blue);

    $\begin{array}{r} 8 \\ +2 \\ \hline 10 \end{array}$ (green); $5 + 5 = 10$ (green); $\begin{array}{r} 6 \\ +9 \\ \hline 15 \end{array}$ (red);
    $4 + 6 = 10$ (green)

2. Sample answer: Ⓓ Ⓓ Ⓓ Ⓝ Ⓟ Ⓟ

Home Link 6·2

1. $9 + 1$, $1 + 9$, $8 + 2$, $2 + 8$, $3 + 7$, $7 + 3$, $6 + 4$, $4 + 6$, $5 + 5$, $10 + 0$, $0 + 10$

2. All names should be equal to 15.

3. All names should be equal to 18.

Home Link 6·3

1. 7, 5, 12
    $7 + 5 = 12$ $5 + 7 = 12$
    $12 - 7 = 5$ $12 - 5 = 7$

2. 6, 9, 15
    $6 + 9 = 15$ $9 + 6 = 15$
    $15 - 6 = 9$ $15 - 9 = 6$

3. 30; 24

Home Link 6·4

Your child should practice addition and subtraction facts using Fact Triangles.

Home Link 6·5

1. 9, 3, 12

  $9 + 3 = 12$ $3 + 9 = 12$

  $12 - 9 = 3$ $12 - 3 = 9$

2. All names should be equal to 14.

3. Your child should cross out $5 + 5 + 5$, $2 + 10$, and tally marks totaling 10.

4. Sample answer: | | | | | ● ●

Home Link 6·6

1-4 The lengths recorded should match the lengths of the objects chosen by your child.

5. 10   **6.** 9

Home Link 6·7

Your child should practice addition and subtraction facts using Fact Triangles.

Home Link 6·8

1. Sample answers:

  ⒟⒟⒟⒟ⓅⓅⓅ; ⒟⒟ⓃⓃⓃⓃⓅⓅⓅ

2. Sample answers:

  ⒟⒟⒟⒟⒟⒟ⓃⓅⓅ;

  ⒟⒟ⓃⓃⓃⓃⓃⓃⓃⓃⓃⓃⓅⓅ

4. 9   **5.** 9   **6.** 5

Home Link 6·9

1. 50¢ or $ 0.50   **2.** 82¢ or $ 0.82

3. 43¢ or $ 0.43   **4.** 66¢ or $ 0.66

5. 74; 75; 77

Home Link 6·10

1.   **2.**   **3.**

Home Link 6·11

Your child should practice addition and subtraction facts using Fact Triangles.

Home Link 6·12

1. 25 children   **2.** 18

3. 5   **4.** 13

5. All names should be equal to 12.

HOME LINK 6·1

Finding Addition Sums

Family Note Children continue practicing addition facts. Today they learned how to use the facts table to find sums. Have your child explain how the table shows the fact 6 + 8 = 14.

Help your child find sums in the table. It is fine to solve the problems using other strategies, such as counting on or using counters to model the problems.

Please return page 163 to school tomorrow. Keep the facts table at home for future use.

		8	9
		8	9
		9	10
		10	11
		11	12
		12	13
		13	14

6	6	7	8	9	10	11	12	13	14	15
7	7	8	9	10	11	12	13	14	15	16
8	8	9	10	11	12	13	14	15	16	17
9	9	10	11	12	13	14	15	16	17	18

+, −	0	1	2	3	4	5	6	7	8	9
0	0	1	2	3	4	5	6	7	8	9
1	1	2	3	4	5	6	7	8	9	10
2	2	3	4	5	6	7	8	9	10	11
3	3	4	5	6	7	8	9	10	11	12
4	4	5	6	7	8	9	10	11	12	13
5	5	6	7	8	9	10	11	12	13	14
6	6	7	8	9	10	11	12	13	14	15
7	7	8	9	10	11	12	13	14	15	16
8	8	9	10	11	12	13	14	15	16	17
9	9	10	11	12	13	14	15	16	17	18

HOME LINK 6·1

Finding Addition Sums *continued*

Use the color code to color the picture.

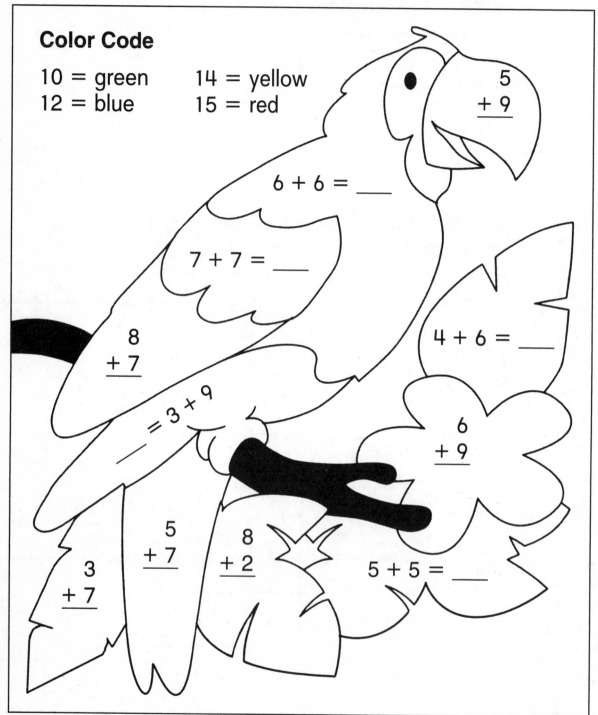

Color Code

10 = green   14 = yellow
12 = blue   15 = red

5
+ 9
___

6 + 6 = ___

7 + 7 = ___

8
+ 7
___

4 + 6 = ___

___ = 3 + 9

6
+ 9
___

5
+ 7
___

8
+ 2
___

3
+ 7
___

5 + 5 = ___

___

Practice

Show 37¢. Use Ⓓ, Ⓝ, and Ⓟ. _____

Name-Collection Boxes

> **Family Note** Today we began working with name-collection boxes. See the attached letter for more information about this routine.
>
> *Please return this Home Link to school tomorrow.*

1. List all of the addition facts you know that have a sum of 10.

_____

_____

2. Write as many names as you can in the name-collection boxes.

15	18
10 + 5	

Practice

3. How old were you 2 years ago? _____

4. Odd or even? _____

HOME LINK 6·2 Family Letter

Name-Collection Boxes

People, things, and ideas often have several different names. For example, Mary calls her parents Mom and Dad. Other people may call them Linda and John, Aunt Linda and Uncle John, or Grandma and Grandpa. Mail may come addressed to Mr. and Mrs. West. All of these names are for the same two people.

Your child is bringing home an activity with a special format for using this naming idea with numbers. We call this format a name-collection box. The box is used by children to collect many names for a given number.

The box is identified by the name on the label. The box shown here is a 25-box, a name-collection box for the number 25.

Names can include sums, differences, products, quotients, or combinations of operations, as well as words (including words in other languages), tally marks, and arrays. A name-collection box can be filled by using any equivalent names.

With repeated practice, children gain the power to rename numbers for a variety of different uses.

25	
37 − 12	20 + 5
卌 卌 卌 卌 卌	
twenty-five	
veinticinco	x x x x x
	x x x x x
	x x x x x
	x x x x x
	x x x x x

HOME LINK 6·3 | # Fact Families

Family Note   We have extended our work with facts to subtraction facts by introducing fact families. Your child will generate addition facts and subtraction facts for the numbers pictured on the dominoes below.

Note that for each problem, there are two addition facts and two subtraction facts.

Please return this Home Link to school tomorrow.

Write the 3 numbers for each domino. Use the numbers to write the fact family.

1.

Numbers: _____, _____, _____

Fact family:

_____ + _____ = _____

_____ + _____ = _____

_____ − _____ = _____

_____ − _____ = _____

2.

Numbers: _____, _____, _____

Fact family:

_____ + _____ = _____

_____ + _____ = _____

_____ − _____ = _____

_____ − _____ = _____

Practice

3. Write the missing numbers.

 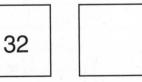

Rule
−2

32 | | 28 | 26 |

147

Family Letter

Fact Triangles

Your child should cut apart the triangles on page 172. Notice that each triangle has the three numbers used in a fact family on it. Use these triangles like flash cards to practice addition and subtraction facts.

The number below the dot is the sum of the other two numbers.
For example, 8 is the sum of 5 and 3.

To practice addition, cover the sum. Your child then adds the numbers that are not covered. For example, if you cover 8, your child adds 5 and 3.

To practice subtraction, cover one of the numbers at the bottom of the triangle. Your child then subtracts the uncovered number at the bottom from the sum. For example, if you cover 3, your child subtracts 5 from 8. If you cover 5, your child subtracts 3 from 8.

Fact Triangles have two advantages over regular flash cards.

1. They reinforce the strong link between addition and subtraction.

2. They help simplify the memorizing task by linking four facts together. Knowing a single fact means that you really know four facts.

$$5 + 3 = 8$$
$$3 + 5 = 8$$
$$8 - 5 = 3$$
$$8 - 3 = 5$$

Save this set of Fact Triangles in an envelope or a plastic bag to continue practicing addition and subtraction facts with your child when you have time.

Fact Triangles

Cut out the 6 triangles. Practice the addition and subtraction facts on these triangles with someone at home.

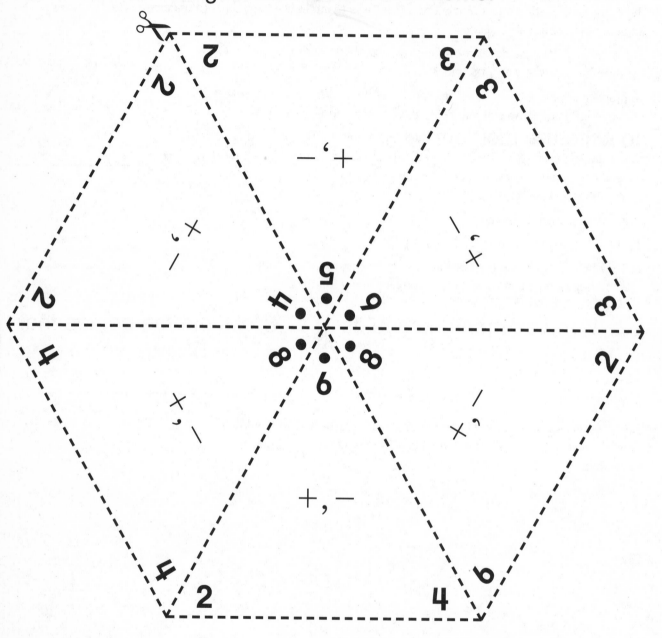

Fact Routines Practice

HOME LINK
6·5

Family Note This Home Link reviews some of the work children have been doing in recent lessons. Note that children are now working with subtraction facts as they are related to addition facts. Encourage your child to include some subtraction "names" in the name-collection box in Problem 2. An example of a subtraction name for 14 is 16 − 2.

Please return this Home Link to school tomorrow.

Write the 3 numbers for the domino. Use the numbers to write the fact family.

1. Numbers: _____, _____, _____

Fact family: ____ + ____ = ____   ____ − ____ = ____

____ + ____ = ____   ____ − ____ = ____

---

2. Write as many names as you can for 14.

14

3. Cross out the names that do not belong.

20

10 + 10

H̶H̶H̶ H̶H̶H̶   5 + 5 + 5

2 + 10   24 − 4

20 + 0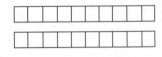

Practice

4. Use | and • to show the number 52.

HOME LINK 6·6 **Measuring in Centimeters**

Family Note Children are beginning to use metric units to measure length (in addition to the U.S. customary units of inches and feet). Your child should measure objects to the nearest centimeter. Make sure your child lines up one end of the object being measured with the "0" mark on the ruler.

Please return this Home Link to school tomorrow.

Find four small objects. Draw a picture of each object. Use your ruler to measure each object to the nearest centimeter (cm). Record your measurements.

1. About _____ cm long	**2.** About _____ cm long
3. About _____ cm long	**4.** About _____ cm long

Practice

Find the total number of dots on the dice.

5. [dice: 6] + [dice: 4] = _____   **6.** [dice: 5] + [dice: 4] = _____

153

HOME LINK 6·7 Practicing with Fact Triangles

Family Note Six more Fact Triangles are being added for practice at home. As you help your child practice, keep the facts your child knows in a separate pile from the facts that still need some work.

Please return this Home Link to school tomorrow.

Cut out the Fact Triangles.
Practice these facts at home.

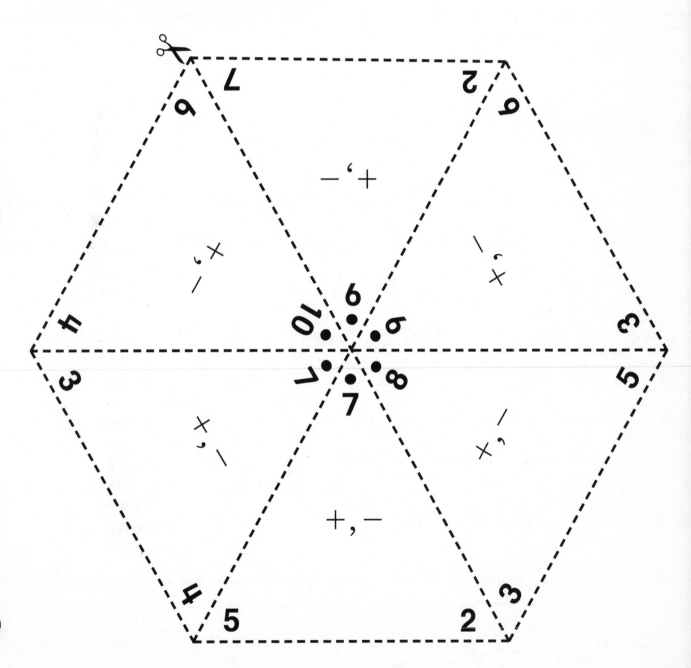

Counting Coins

HOME LINK
6·8

Family Note This Home Link reviews finding the value of combinations of dimes, nickels, and pennies. If your child is having trouble finding the value of collections of coins, you might try the following method, using real coins, if possible:

1. Show the amount with pennies.

2. Trade the pennies for nickels.

3. Trade the nickels for dimes.

Beginning tomorrow, children will add quarters to their work with coins. In preparation, please give your child two quarters to bring to school.

Please return this Home Link to school tomorrow.

Use Ⓟ, Ⓝ, and Ⓓ to show each amount in two different ways.

1. 43¢

2. 67¢

3. Ask someone at home for two quarters. Bring them to school.

Practice

Find the total number of dots for each one.

4.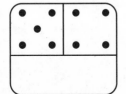

5.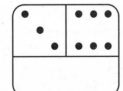

6.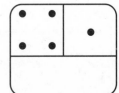

Name _____ Date _____

Family Note Children have begun to find the value of coin combinations that include quarters. If your child is having difficulty because coins are not shown in any particular order, use real coins to model the problems. Sort the coins into groups of like coins (all dimes together, all nickels together) before counting.

Children also continue to use dollars-and-cents notation (for example, $1.05). If your child has trouble recording amounts in this notation, don't worry—this is a skill we will continue to practice throughout the year.

Please return this Home Link to school tomorrow.

Ⓟ	Ⓝ	Ⓓ	Ⓠ
1¢	5¢	10¢	25¢
$0.01	$0.05	$0.10	$0.25

Find the value of the coins.

Write the total in cents and in dollars-and-cents notation.

1. Ⓝ Ⓠ Ⓓ Ⓝ Ⓝ _____ ¢ or $ _____

2. Ⓠ Ⓠ Ⓓ Ⓝ Ⓓ Ⓝ Ⓟ Ⓟ _____ ¢ or $ _____

3. Ⓓ Ⓟ Ⓟ Ⓝ Ⓟ Ⓠ _____ ¢ or $ _____

4. Ⓓ Ⓝ Ⓟ Ⓠ Ⓠ _____ ¢ or $ _____

Practice

5. Fill in the blanks.

72   73   _____   _____   76   _____

159

HOME LINK
6·10

Time at 5-Minute Intervals

Family Note In today's lesson, children started to work with digital displays of time. Children talked about the number of minutes in an hour and started to tell time at 5-minute intervals. This will require a lot of practice, so the *Everyday Mathematics* program will come back to telling time throughout the year.

Please return this Home Link to school tomorrow.

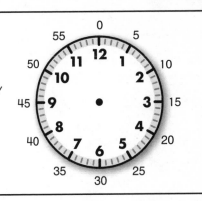

Draw the hour hand and the minute hand.

1.

4:00

2.

7:30

3.

10:15

Practice

4. Draw dots on the domino. Write an addition fact for the domino.

_____ + _____ = _____

HOME LINK 6·11 More Fact Triangles

Continue practicing all addition and subtraction facts.

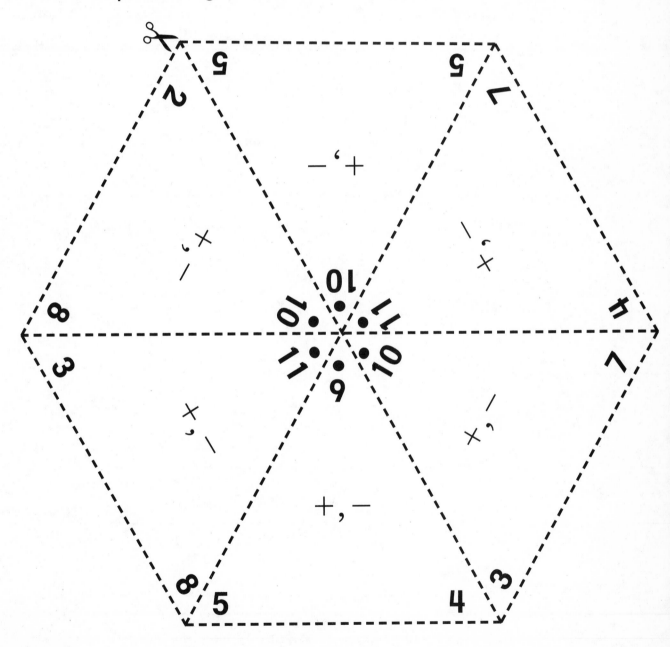

HOME LINK 6·12 Analyzing a Set of Data

Family Note Today we did some calculator counts in class. Ask your child what his or her highest count was at the end of 15 seconds.

Below is a tally chart like one we made in class today. Help your child identify how many children did the counts and the lowest and the highest counts that someone in Casey's class got. Then help your child find the range of the counts. (To find the range, subtract the lowest count from the highest count.)

Please return this Home Link to school tomorrow.

Casey's Class Data for Calculator Counts

Counted to	Number of Children
5	/
7	//
10	////
11	//// /
12	////
13	///
15	//
17	/
18	/

1. How many children in Casey's class did the calculator counts?

_____

2. Find the highest count.

_____

3. Find the lowest count.

_____

4. Find the range of the counts.

_____

Practice

5. Write some names for 12.

Unit 7: Family Letter

Geometry and Attributes

In Unit 7, children will work with 2-dimensional shapes. First, children will classify blocks by their shape, color, and size. Then they will learn to recognize attributes such as number of sides and square corners. Later they will build their own shapes out of straws and twist-ties, identifying the differences among shapes that are polygons and shapes that are not.

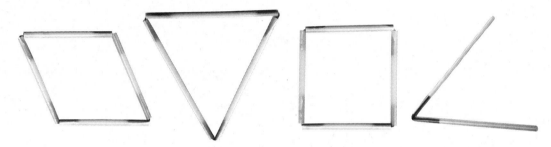

Children will work with 3-dimensional shapes they encounter every day. They will be asked to bring objects from home, which will be organized and labeled to create a "Shapes Museum" for the classroom. For example, a soup can would be labeled "cylinder"; a tennis ball, "sphere." In examining the shapes brought to class, children will begin to identify similarities and differences among five kinds of 3-dimensional shapes: prisms, pyramids, spheres, cylinders, and cones. They will learn to identify characteristics, using terms such as *flat* and *round*. We will use the names of the shapes in class, but children will not be expected to memorize their definitions.

In the last lesson of this unit, children will explore symmetry. They will find symmetrical shapes in real life, including butterflies, bells, guitars, vases, and double dominoes. Then they will create their own symmetrical shapes, using paper and scissors.

Please keep this Family Letter for reference as your child works through Unit 7.

Vocabulary

2-Dimensional Shapes

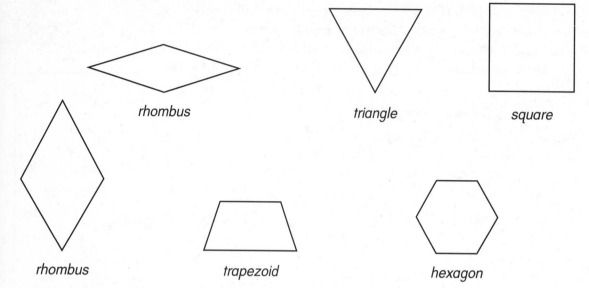

rhombus

triangle

square

rhombus

trapezoid

hexagon

3-Dimensional Shapes

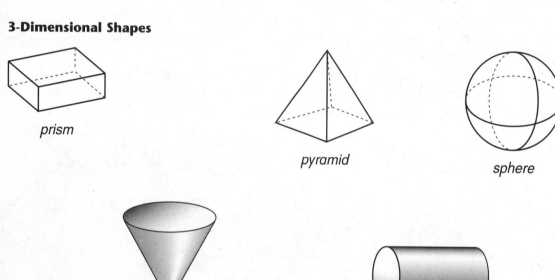

prism

pyramid

sphere

cone

cylinder

Do-Anytime Activities

To work with your child on the concepts taught in this unit and in previous units, try these interesting and rewarding activities:

1. Draw a name-collection box with a number on the tag. Ask your child to write at least 10 equivalent names for the given number.

2. Occasionally ask questions about time: What time is it? What time will it be in five minutes? In ten minutes? In one hour?

3. Continue to work on addition and subtraction facts using Fact Triangles, short drill sessions, and any of the games introduced at school.

4. If a calculator is available, ask your child to show you how to count with it. See how high your child can count on the calculator.

5. Look for geometric shapes around the house, at the supermarket, as part of architectural features, and on street signs. Begin to call these shapes by their geometric names.

12
$17 - 5$
$2 + 10$
$4 + 8$
$13 - 1$
twelve
doce
$\cancel{||||}\ \cancel{||||}\ ||$

Building Skills through Games

In this unit, your child will practice classification and place-value skills by playing the following games:

Attribute Train Game

One player puts down a block. The next player finds a block that differs in only one attribute—shape, size, or color—from the first block and puts it next to the first block. Each player continues to add to the "train" of blocks.

Tens-and-Ones Trading Game

Players take turns putting base-10 blocks on their Tens-and-Ones Mat according to the roll of a die. Whenever possible, they exchange 10 cubes for 1 long. The first player to get 10 longs wins!

As You Help Your Child with Homework

As your child brings assignments home, you may want to go over the instructions together, clarifying them as necessary. The answers listed below will guide you through the Home Links in this unit.

Home Link 7·1

Check that your child answers facts correctly as he or she practices with the Fact Triangles.

Home Link 7·2

1.–4. Answers vary.

5. Kente; 5¢

Home Link 7·3

1. square, rhombus, hexagon; trapezoid, triangle, rhombus

2. $6 + 1 = 7$; $7 - 1 = 6$; $1 + 6 = 7$; $7 - 6 = 1$

Home Link 7·4

1.

2. Sample drawing:

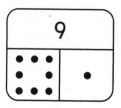

3. Sample drawing:

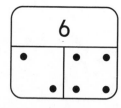

Home Link 7·5

Check that your child answers facts correctly as he or she practices with the Fact Triangles.

Home Link 7·6

1. Answers vary.

2. 71; 72; 74

Home Link 7·7

1. Answers vary.

2. 11

Practicing with Fact Triangles

Family Note Your child should cut apart the Fact Triangles below. Add these to the Fact Triangles from earlier lessons. As you help your child practice facts, separate the triangles into piles to show the facts your child knows and the facts that still need work. Continue to practice all of the facts.

Cut out these Fact Triangles. Practice the facts at home.

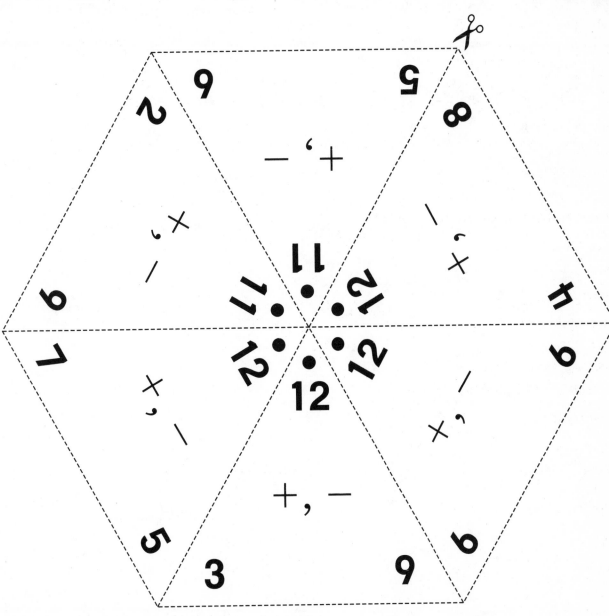

Shapes

> **Family Note** The class has been identifying and comparing three attributes of objects: shape, color, and size. We will work with 2-dimensional and 3-dimensional shapes in future lessons. To prepare for this, help your child find objects with the shapes listed below. Also help your child find objects to bring to school for our Shapes Museum. The objects should not be valuable or breakable.
>
> *Please return this Home Link to school tomorrow.*

1. Find something in your house that has a triangle in it. Write its name or draw its picture.

2. Find something in your house that has a circle in it. Write its name or draw its picture.

3. Find something in your house that has a square in it. Write its name or draw its picture.

4. Starting tomorrow, bring things to school for the Shapes Museum.

Practice

5. Kente has ⓓ ⓓ ⓓ ⓝ ⓟ.

Rossita has ⓓ ⓓ ⓝ ⓝ ⓟ.

Who has more money? _____

How much more money? _____ ¢

HOME LINK 7·3 | **Polygons**

Family Note We are beginning to identify polygons and their characteristics. A polygon is a closed 2-dimensional figure. It is formed by three or more line segments that meet only at their endpoints.

On this page, your child will try to name the shapes we worked with today. Some of the names may still be confusing.

Please return this Home Link to school tomorrow.

1. Use the Word List to help you write the name of each shape.

Word List

hexagon   rhombus   square   trapezoid   triangle

_____   _____   _____

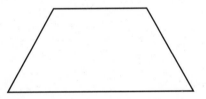   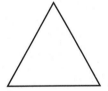   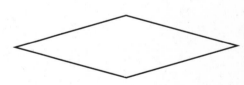

_____   _____   _____

Practice

2. Write the fact family for this domino.

_____ + _____ = _____   _____ − _____ = _____

_____ + _____ = _____   _____ − _____ = _____

175

HOME LINK 7·4 | Identifying Polygons

MRB 52–53

Family Note This Home Link follows up our work in class identifying shapes called polygons. A polygon is a closed 2-dimensional figure formed by three or more line segments that meet only at their endpoints. Some examples of polygons are shown below.

Help your child identify the polygons in Problem 1.

 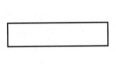 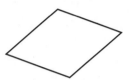 

Please return this Home Link to school tomorrow.

1. Circle the 5 polygons.

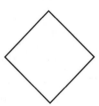    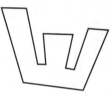

Practice

Draw the missing dots.

2.
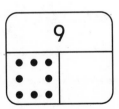

3.

Practicing with Fact Triangles

Family Note Your child should cut apart the Fact Triangles below. Add these to the Fact Triangles from earlier lessons. As you help your child practice facts, separate the triangles into piles to show the facts that your child knows and the facts that still need work. Continue to practice all of the facts.

Continue practicing all of the addition and subtraction facts at home.

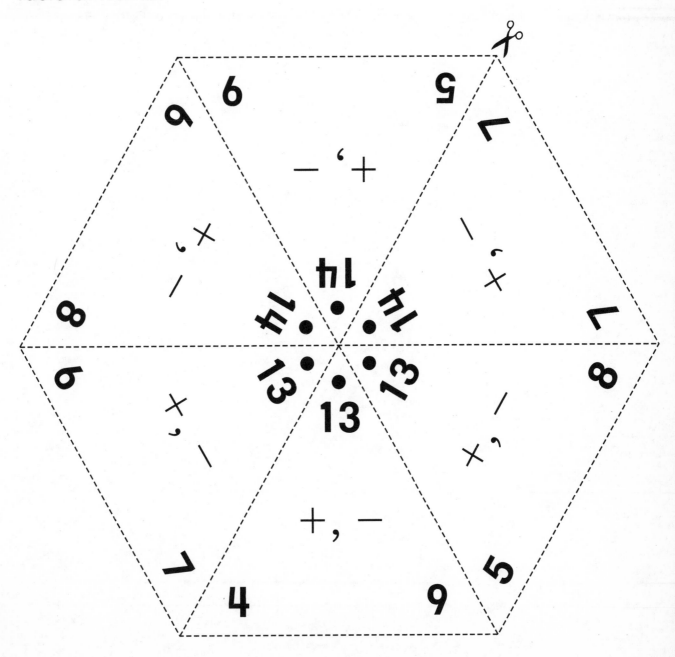

HOME LINK 7·6 Tracing Shapes

1. Find 3-dimensional shapes with flat faces (sides).

On the back of this page, trace around one face of each shape.

Write the name of the shape on each tracing.

Word List		
square	circle	hexagon
trapezoid	rhombus	triangle
not a polygon	rectangle	other polygon

Practice

2. Fill in the blanks.

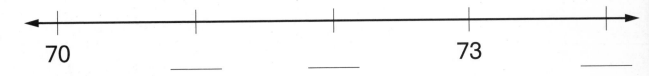

70 ___ ___ 73 ___

HOME LINK 7·7 | **Finding Symmetry in Nature**

> **Family Note** A picture or an object has symmetry if it can be folded in half so that the two halves match exactly. In today's lesson, the class explored symmetry by cutting out designs from folded paper.
>
> To continue our exploration of symmetry, help your child find pictures that show symmetry in nature; for example, pictures of butterflies, leaves, animal markings, flowers, or snowflakes.
>
> *Please return this Home Link to school tomorrow.*

1. Find symmetrical pictures in magazines.

Cut out your favorite pictures and glue them onto this page.

Practice

2. Record the time.

quarter-to _____ o'clock

HOME LINK 7·8

Unit 8: Family Letter

Mental Arithmetic, Money, and Fractions

In Unit 8, children will examine a dollar bill and add the dollar to the money units they already know. They will continue to count and record amounts of money (using pennies, nickels, dimes, and quarters), often in more than one way. They will also begin learning how to make change.

Children will also create addition, subtraction, and comparison problems for the class to solve and will share their own problem-solving strategies. Having children share their solution strategies is emphasized in *Everyday Mathematics* and helps children feel more confident as they express their ideas.

Later in Unit 8, children will work with fractions. They will be reminded that fractions are equal parts of wholes. When dealing with fractions, it is important that children keep in mind the "whole" or the ONE to which the fraction is linked. For example, $\frac{1}{2}$ of an apple and $\frac{1}{2}$ of a dollar are not the same because they deal with different types of "wholes."

Please keep this Family Letter for reference as your child works through Unit 8.

Vocabulary

Important terms in Unit 8:

fractional parts Equal parts of any whole.
For example:

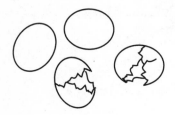

Half ($\frac{1}{2}$) of the whole set of 4 eggs
are broken.

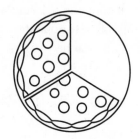

One-third ($\frac{1}{3}$) of a whole
3-slice pizza has been eaten.

near doubles A strategy derived from the "doubles addition
facts." For example, a child might solve 3 + 4 by noting that
3 + 3 = 6, so 3 + 4 must be 1 more than 6, or 7.

Do-Anytime Activities

To work with your child on the concepts taught in this unit and in previous units, try these interesting
and rewarding activities:

1. Continue to review addition and subtraction facts.

2. Ask questions like the following:

♦ I want to buy an airplane that costs 27 cents. If I give the clerk
3 dimes, how much change will I get back?

♦ How can you show 14 cents using exactly 6 coins? (Have the
actual coins available.)

♦ How many different ways can you show 14 cents? (Have the
actual coins available.)

3. Count out 8 pennies (or any type of counter, such as buttons or
paper clips). Ask your child to show you $\frac{1}{2}$ of the pennies and
then $\frac{1}{4}$ of the pennies. Do this with a variety of different numbers.

4. Encourage your child to count various collections of coins you may
have accumulated.

Building Skills through Games

In Unit 8, your child will practice addition, subtraction, place value, and money skills by playing the following games:

Addition Top-It

See *My Reference Book*, pages 122–123. Players turn over two cards and call out the sum. The player with the higher sum keeps all the cards. The player with more cards at the end wins.

Base-10 Exchange

Players roll the dice and put that number of cubes on their Place-Value Mats. Whenever possible, they exchange 10 cubes for 1 long. The first player to make an exchange for a flat wins.

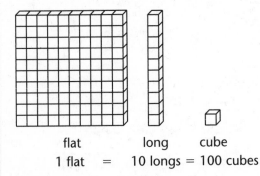

flat   long   cube
1 flat   =   10 longs = 100 cubes

One-Dollar Exchange

See *My Reference Book*, pages 144–145. Players roll the dice and put that number of cents on their Place-Value Mats. Whenever possible, they exchange 10 pennies for 1 dime. The first player to make an exchange for a $1 bill wins.

3, 2, 1, Game

See *My Reference Book*, pages 150–151. Players take turns subtracting 1, 2, or 3 from a given number. The first player to reach 0 exactly is the winner.

As You Help Your Child with Homework

As your child brings home assignments, you may want to go over the instructions together, clarifying them as necessary. The answers listed below will guide you through the Home Links in this unit.

Home Link 8·1

1. Sample answer: Your child should mark 2 dimes, 3 nickels, and 2 pennies.

2. Your child should mark 1 quarter, 4 dimes, and 1 nickel.

3. 52   **4.** 61   **5.** 96   **6.** 88

7. <   **8.** >   **9.** =

Home Link 8·2

1. Sample answer:

2. Sample answer:

3. 111¢, $1.11; Ⓠ Ⓠ Ⓠ Ⓠ Ⓓ Ⓟ

4. 8, even

Home Link 8·3

1. 569   **2.** 483   **3.** 709   **4.** Grant; 9¢

Home Link 8·4

1. Your child should tape or glue a picture to the page or back of the page, tell a number story, and write a number model that goes with his or her story.

2. 12   **3.** 9   **4.** 11   **5.** 13
6. 10   **7.** 12

Home Link 8·5

1. 3, 4   **2.** 1, 5   **3.** +5; 20, 25

Home Link 8·6

1. Sample answer:

2. Sample answer:

3. 18, 19, 22

Home Link 8·7

1.   **2.**

3.   **4.** 453

Home Link 8·8

1. ; 5

2.

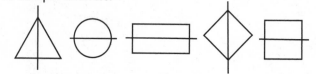; 4

3. 4

4. Sample answer: 26, even

Home Link 8·9

1. 4 + 6 = 10   **2.** 9 + 1 = 10   **3.** 9 + 9 = 18
4. 7 + 3 = 10   **5.** 4 + 4 = 8   **6.** 6 + 1 = 7
    3 + 7 = 10   8 − 4 = 4   1 + 6 = 7
    10 − 3 = 7   7 − 6 = 1
    10 − 7 = 3   7 − 1 = 6

7. Sample answers:

Coin Combinations

> **Family Note** In the next lesson, we will extend our work with money to include dollars. In preparation for this, we have been practicing counting coins. If your child has difficulty with some problems on this page, use real coins to model the situations. Arrange the coins in groups of like coins and count the coins of the highest value first.
>
> *Please return this Home Link to school tomorrow.*

1. Mark the coins you need to buy an eraser.

2. Mark the coins you need to buy a box of crayons.

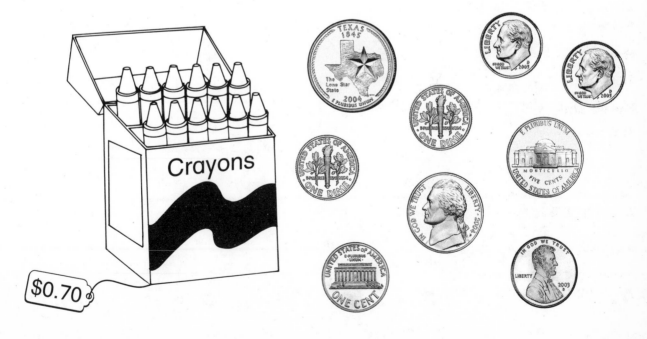

Coin Combinations *continued*

Martina saved her money.
How much did she save each month?

3. September Total: ___43___ ¢

4. October   Total: ___61___ ¢

5. November Total: ___96___ ¢

6. December Total: ___88___ ¢

Practice

Write <, >, or =.

7. 13 ⟨<⟩ 42   **8.** 106 ⟨>⟩ 105   **9.** 4 + 5 ⟨=⟩ 9

HOME LINK 8·2 Dollars and More ✓

Family Note Today we took a close look at a dollar bill. Since we have only begun to work with dollars, some of the problems on this page may be difficult for your child. If possible, use real money to model the problems. Start by counting the bills and coins in the example with your child.

Please return this Home Link to school tomorrow.

Show how you would pay for each item.
Use $1, Q, D, N, or P.

223¢   100 + 11
$2.23 = 111

Example:

$1.95   MYSTERY

1.

$2.85

$1 $1
Q Q Q D D
1¢

2.

$3.24

99
100

10 + 1
1
2
463
$.63

$1 Q Q Q D D

3. Write this amount in two ways.

Q Q N D N D P D D D

Total: _III____ ¢ $ _1.11____

Show this amount using fewer coins.

199
200

_____

Practice

4. Circle the tens place. Is the number odd or even?

8(6) even

HOME LINK 8·3 More Riddles

Family Note   We are extending our work with base-10 blocks to include 100s. The base-10 block for 100 is called a "flat." Note that the blocks are not always shown in the same order. If your child finds some of the problems difficult, you might model them with dollar bills (for flats), dimes (for longs), and pennies (for cubes). These make good substitutes for base-10 blocks.

Ask your child to explain why there is a zero in the number in Problem 3. To practice reading 3-digit numbers, ask your child to read his or her answers to you.

Please return this Home Link to school tomorrow.

Hundreds	Tens	Ones

Solve the riddles.

Example:

2 [flat]   5 [long]   7 [cube]   What am I? __257__

1. 5 [flat]   6 [long]   9 [cube]   What am I? _____

2. 4 [flat]   8 [long]   3 [cube]   What am I? _____

3. 7 hundreds and 9 ones   What am I? _____

Practice

4. Grant has Ⓠ Ⓠ Ⓠ Ⓓ Ⓝ. Joanna has Ⓠ Ⓠ Ⓠ Ⓝ Ⓟ.

Who has more money? _____.

How much more money? _____ ¢

A Shopping Story

HOME LINK 8·4

> **Family Note** We have been practicing addition of 2-digit numbers using number stories about money. Please help your child find pictures of two items in a magazine, newspaper, or catalog that each cost less than one dollar. (Newspaper inserts tend to be a good source for such items.) Ask your child to make up and tell you a number story to go with the items.
>
> *Please return this Home Link to school tomorrow.*

Sample Story

I bought a ball and an eraser. I paid 52 cents.

Number model $35¢ + 17¢ = 52¢$

1. Glue or tape your pictures below or on the back of this page. Write your story.

Number model: _____

Practice

Find the sums.

2. $\begin{array}{r} 6 \\ + 6 \\ \hline \end{array}$ **3.** $\begin{array}{r} 5 \\ + 4 \\ \hline \end{array}$ **4.** $\begin{array}{r} 10 \\ + 1 \\ \hline \end{array}$ **5.** $\begin{array}{r} 9 \\ + 4 \\ \hline \end{array}$

6. $1 + 9 =$ _____   **7.** $10 + 2 =$ _____

HOME LINK 8·5 | **Making Change**

> **Family Note** Children are beginning to learn how to make change. If you have dimes, nickels, and pennies available, have your child act out the problems with real money. For each problem, your child should pay with just enough dimes to cover the cost.
>
> *Please return this Home Link to school tomorrow.*

Record the number of dimes you paid.
Record the amount of change you got.

Example:	**1.**	**2.**
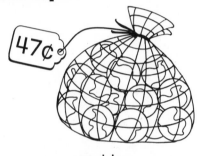		
marbles	balloon	toy car
I paid with __5__ dimes.	I paid with ____ dimes.	I paid with ____ dime.
I got __3__ cents in change.	I got ____ cents in change.	I got ____ cents in change.

Practice

3. Find the rule. Write the missing numbers.

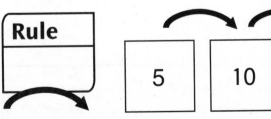

Rule
_____

| 5 | 10 | 15 | | |

HOME LINK 8·6 | **Exploring Halves and Fourths**

> **Family Note** We are beginning to explore the concept of fractions. Today, children focused on identifying fractional parts of things. We emphasized that fractional parts come from dividing something into equal parts.
>
> *Please return this Home Link to school tomorrow.*

1. Divide each of the squares in half. Try to divide each square in a different way.

2. Divide each of the squares into fourths. Try to divide each square in a different way.

 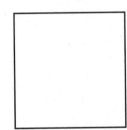 

Practice

3. Fill in the blanks.

_____   _____   20   21   _____   23

Equal Parts

> **Family Note** Ask your child to explain how he or she knows which figures are divided into equal fractional parts. Then help your child write fractions in the equal parts.
>
> *Please return this Home Link to school tomorrow.*

Circle each shape that shows equal parts.
Write fractions in the equal parts.

1.

halves

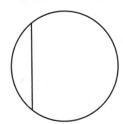 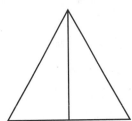 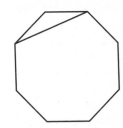

2.

sixths

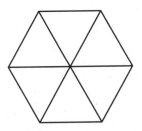 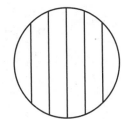 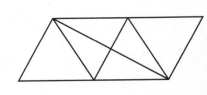

3.

fourths

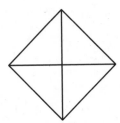 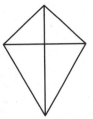 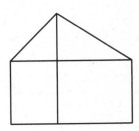

Practice

4. Write a 3-digit number with 4 in the hundreds place,

5 in the tens place, and 3 in the ones place. _____

Sharing Sets of Objects

Family Note Today we extended our work with fractions to finding fractional parts of collections of objects. Help your child act out the problems below with pennies or counters. When sharing things equally, one strategy is to distribute the things just as you would deal cards in a card game and then count the things in one share.

Please return this Home Link to school tomorrow.

Use pennies to help you solve the problems.

1. Halves: 2 people share 10 pennies equally.
Circle each person's share.

ⓟ ⓟ ⓟ ⓟ ⓟ ⓟ ⓟ ⓟ ⓟ ⓟ

How many pennies does each person get? _____ pennies

2. Thirds: 3 children share 12 balloons equally.
Draw the balloons that each child gets.

How many balloons does each child get? _____ balloons

3. Fourths: 4 children share 16 flowers equally.
How many flowers does each child get? _____ flowers

Practice

4. How old will you be in 20 years? _____

Is the number odd or even? _____

HOME LINK 8·9 | Facts Practice

Fill in the missing numbers.

1.

____ + 6 = ____

2.

____ + 1 = ____

3.

____ + ____ = ____

Write the fact family for each triangle below.

4.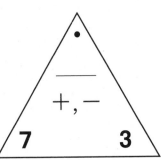

____ + ____ = ____

____ + ____ = ____

____ − ____ = ____

____ − ____ = ____

5.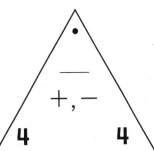

____ + ____ = ____

____ − ____ = ____

6.

7

+, −

6   ____

____ + ____ = ____

____ + ____ = ____

____ − ____ = ____

____ − ____ = ____

Practice

7. Draw a line to divide each shape in half.

  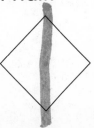 

205

HOME LINK 8·10 | Unit 9: Family Letter

Place Value and Fractions

The concept of place value (ones, tens, hundreds, and so on) that children have worked on since *Kindergarten Everyday Mathematics* will be taught on a more formal level in this unit. Patterns on number grids will be used to reinforce place-value concepts. For example, children may be asked to identify a hidden number on the number grid and to describe the strategies used to find and name that number. Once they are able to do this, they will solve number-grid puzzles— pieces of a number grid with all but a few numbers missing. Here are a few examples of number-grid puzzles:

Children know that all numbers are written with one or more of these 10 digits: 0, 1, 2, 3, 4, 5, 6, 7, 8, and 9. In order to reinforce this understanding, children will identify the place value of different digits in 2- and 3-digit numbers. Help your child remember that these same digits are also used to express quantities less than 1 with fractions.

Later in this unit, children will extend their understanding of fraction concepts as they see relationships among fraction words, meanings, and symbols.

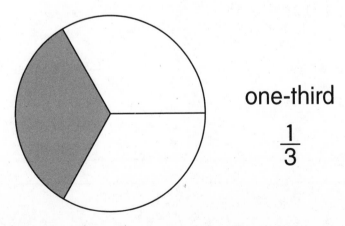

one-third

$\dfrac{1}{3}$

Please keep this Family Letter for reference as your child works through Unit 9.

Vocabulary

Important terms in Unit 9:

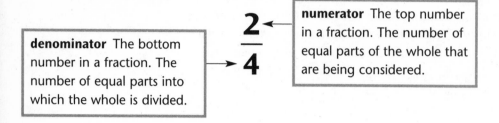

denominator The bottom number in a fraction. The number of equal parts into which the whole is divided.

$\dfrac{2}{4}$

numerator The top number in a fraction. The number of equal parts of the whole that are being considered.

2-digit numbers In base 10, numbers from 10 through 99 that have two digits each.

3-digit numbers In base 10, numbers from 100 through 999 that have three digits each.

Do-Anytime Activities

To work with your child on concepts taught in this unit and in previous units, try these interesting and rewarding activities:

1. Ask questions, such as the following: *What is the fraction word for each of 4 equal parts of something?* (fourths) *Each of eight equal parts?* (eighths)

2. Give your child several pieces of paper to fold into halves, fourths, or eighths. He or she can label each part with the appropriate fraction symbol ($\frac{1}{2}, \frac{1}{4}, \frac{1}{8}$).

3. Using a set of numbers, have your child write the largest and smallest 2- and 3-digit whole numbers possible. For example, using 5, 2, and 9, the largest whole number is 952; the smallest is 259.

4. Say a 2- or 3-digit number. Then have your child identify the actual value of the digit in each place. For example, in the number 952, the value of the 9 is 900, the value of the 5 is 50, and the value of the 2 is 2 ones, or two. An important goal of *Everyday Mathematics* is for children eventually to think of any digit in a multidigit number by its place-value name.

Building Skills through Games

In Unit 9, your child will practice addition skills by playing the following games:

Number-Grid Game See *My Reference Book,* pages 142–143. Each player rolls a die and moves his or her marker on the number grid. The first player to get to 110 or past 110 wins.

Fact Power Game Players take turns rolling a die and moving their markers on the game mat. Players then say the sum for the addition fact on the game mat.

As You Help Your Child with Homework

As your child brings assignments home, you may want to go over the instructions together, clarifying them as necessary. The answers listed below will guide you through the Home Links in this unit.

Home Link 9·1

1. Your child should complete the number grid from 101–200.

2. 269; 272; 273

Home Link 9·2

1. 41 **2.** 71 **3.** 23 **4.** 72 **5.** 78

6. 66 **7.** 65 **8.** 79 **9.** 38

10. 31 **11.** 50

Home Link 9·3

1. 43, 63, 73, 83

2. 24, 25 (across); 33, 53, 63, 73 (down); 64 (across)

3. 59, 69, 89 (down); 78, 80 (across); 88, 90 (across)

4. Sample answers: square, rectangle, rhombus, trapezoid

Home Link 9·4

1. | | | | | | | . . . . ; 77

2. | | | | | . . . . ; 58

3. 71 **4.** 75 **5.** 59

6. 20

Home Link 9·5

1.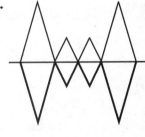

2.

3.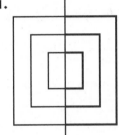

4.

5.

6. no

7. no

8. no

Home Link 9·6

1. $\frac{1}{5}$ **2.** $\frac{2}{3}$ **3.** $\frac{5}{6}$

4. Sample answer: **5.** Sample answer:

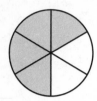 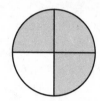

6. Sample answer:

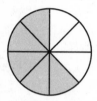

7. Possible answers: window, table, pillow, picture frame

Home Link 9·7

1. Sample answer: A; the half is larger.

2. $\frac{1}{3}$ **3.** $\frac{1}{4}$

4. $7 + 6 = 13$; $13 - 6 = 7$; $13 - 7 = 6$

Home Link 9·8

1. Sample answer: **2.** Sample answer:

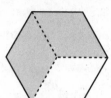 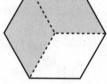

3. Sample answer: **4.** hexagon, square

5. 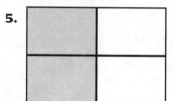 $\frac{2}{4}$

6. 569 **7.** 734

Number-Grid Hunt

Family Note

Ask your child to describe some of the patterns in the number grid below. Then ask him or her to fill in specific numbers you suggest; for example, *Where would the number 140 go?* Do this with several numbers before your child completes the rest of the grid. By learning to identify and use patterns in the number grid, your child will develop strong number sense and computation skills.

Please return this Home Link to school tomorrow.

Ask someone to say a number between 101 and 200. Record it on the number grid. Do this for several numbers. Then finish filling in the grid on your own.

101									
				125					
								139	
									150
171									
		183							

Practice

Count up by 1s.

268, _____, 270, 271, _____, _____, 274

HOME LINK 9·2 # Using the Number Grid

Family Note Ask your child to explain how to count up and back by 10s on the number grid and then to demonstrate how to solve the addition and subtraction problems on the number grid. If your child counts one space at a time, remind him or her that to count up by 10s, you can move down one row for every 10, and to count back by 10s, you can move up one row for every 10.

Please return this Home Link to school tomorrow.

Use the number grid to solve the problems.

−9	−8	−7	−6	−5	−4	−3	−2	−1	0
1	2	3	4	5	6	7	8	9	10
11	12	13	14	15	16	17	18	19	20
21	22	23	24	25	26	27	28	29	30
31	32	33	34	35	36	37	38	39	40
41	42	43	44	45	46	47	48	49	50
51	52	53	54	55	56	57	58	59	60
61	62	63	64	65	66	67	68	69	70
71	72	73	74	75	76	77	78	79	80
81	82	83	84	85	86	87	88	89	90
91	92	93	94	95	96	97	98	99	100

1. $35 + 6 =$ _____

2. $61 + 10 =$ _____

3. $43 - 20 =$ _____

4. _____ $= 82 - 10$ **5.** _____ $= 58 + 20$ **6.** _____ $= 75 - 9$

7.
$$
\begin{array}{r}
55 \\
+ 10 \\
\hline
\end{array}
$$

8.
$$
\begin{array}{r}
99 \\
- 20 \\
\hline
\end{array}
$$

9.
$$
\begin{array}{r}
46 \\
- 8 \\
\hline
\end{array}
$$

Practice

Solve.

10. 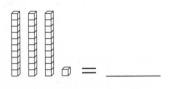 $=$ _____

11. 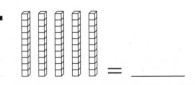 $=$ _____

Number-Grid Puzzles

> **Family Note** Have your child show you how to complete the number-grid puzzles. Encourage him or her to explain patterns on the number grid that are helpful for solving the problems. For example, if you move up one row, the digit in the 10s place is 1 less.
>
> *Please return this Home Link to school tomorrow.*

Show someone at home how to fill in the missing numbers.

1.

53

2.

23	24	25

43

3.

	79	

Practice

4. Draw shapes that have exactly 4 sides and 4 corners.

Write their names.

_____ _____ _____

HOME LINK 9·4 **Solving Problems Two Ways**

Family Note Ask your child to explain how to solve the first set of problems with base-10 blocks and the second set on the number grid. At this point it is important that children work with more concrete representations. This will be beneficial later, when they are faced with more difficult problems.

Please return this Home Link to school tomorrow.

Draw the total number of base-10 blocks.
Then write the total.

Example: |||||.. + |||..... = |||||||||.......
           52 + 35 =   87

1. |..... + ||||||.. = _____

    15 + 62 = _____

2. |||.... + ||.... = _____

    34 + 24 = _____

Use the number grid to help you solve the problems.

3. 63 + 8 = _____

4. 55 + 20 = _____

5. _____ = 47 + 12

−9	−8	−7	−6	−5	−4	−3	−2	−1	0
1	2	3	4	5	6	7	8	9	10
11	12	13	14	15	16	17	18	19	20
21	22	23	24	25	26	27	28	29	30
31	32	33	34	35	36	37	38	39	40
41	42	43	44	45	46	47	48	49	50
51	52	53	54	55	56	57	58	59	60
61	62	63	64	65	66	67	68	69	70
71	72	73	74	75	76	77	78	79	80
81	82	83	84	85	86	87	88	89	90
91	92	93	94	95	96	97	98	99	100

Practice

6. It is 8:10. How many minutes is it until 8:30?

_____ minutes

217

Symmetry

Family Note In class today, children used blocks to make the mirror image of a design across a line of symmetry. This resulted in a symmetrical design. A figure is symmetrical across a line if it has two matching halves. On this page, help your child complete the designs so that they are symmetrical.

Please return this Home Link to school tomorrow.

Complete each design so that the two halves match.

Example:

1.

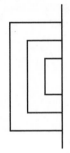

2.

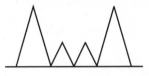

3.

4.

5.

Practice

Yes or no?

6. $0.85 > 85¢ _____

7. 5 pennies < 5¢ _____

8. Ⓝ Ⓟ Ⓟ Ⓟ = 1 dime _____

HOME LINK 9·6   Fractional Parts

Family Note   In Unit 8, we worked with unit fractions, such as $\frac{1}{2}$, $\frac{1}{3}$, and $\frac{1}{4}$. Today, we started to explore fractions in which the number above the fraction bar is more than 1, such as $\frac{2}{3}$, $\frac{3}{4}$, and $\frac{5}{6}$. If your child is having trouble with some of the problems on this page, you might mention that $\frac{1}{2}$ means that 1 out of 2 parts is shaded, that $\frac{3}{6}$ means that 3 out of 6 parts are shaded, and so on. Or you might ask your child to explain the fractions to you in this way.

Please return this Home Link to school tomorrow.

Mark the fraction that tells what part of the circle is shaded.

1.

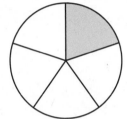

$\frac{1}{2}$ $\frac{5}{6}$

$\frac{1}{5}$ $\frac{5}{1}$

2.

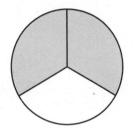

$\frac{2}{2}$ $\frac{2}{3}$

$\frac{3}{4}$ $\frac{3}{1}$

3.

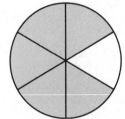

$\frac{1}{6}$ $\frac{1}{5}$

$\frac{6}{5}$ $\frac{5}{6}$

Shade the circles.

4.

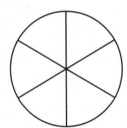

Shade $\frac{4}{6}$.

5.

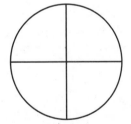

Shade $\frac{3}{4}$.

6.

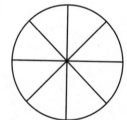

Shade $\frac{5}{8}$.

Practice

7. Name or draw 4 squares you find in your home.

HOME LINK 9·7 Comparing Fractions

Family Note

Today we divided unit strips into equal parts: halves, thirds, fourths, sixths, and eighths. Then we compared the sizes of the parts. Your child probably cannot tell which of two fractions is more by looking at the fractions, but he or she should be able to compare two fractions by looking at pictures of them. Encourage your child to label one part of each shape with a fraction before deciding which fraction is more or less.

Please return this Home Link to school tomorrow.

1. Which would you rather have, half of fruit bar A or half of fruit bar B? Explain your answer to someone at home.

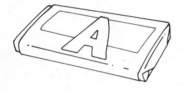

2. Which is more, $\frac{1}{5}$ or $\frac{1}{3}$?

_____

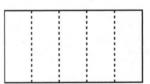

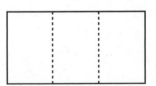

3. Which is more, $\frac{1}{4}$ or $\frac{1}{6}$?

_____

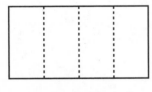

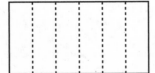

Practice

4. Write the fact family.

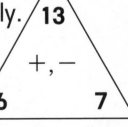

$\underline{}6\underline{} + \underline{}7\underline{} = \underline{}13\underline{}$

$\underline{} + \underline{} = \underline{}$

$\underline{} - \underline{} = \underline{}$

$\underline{} - \underline{} = \underline{}$

Name _____ Date _____

Solving Fraction Problems

Family Note This Home Link reviews some of the fraction concepts we have covered this year. The most important concept first graders should understand is that a fraction names a part of something (the whole) that has been divided into equal parts. Because children's work on fraction concepts this year may be their first exposure, they may still be unclear about some of the ideas we have explored. That's okay; these and other fraction concepts will be revisited in later grades.

Please return this Home Link to school tomorrow.

1. Shade $\frac{1}{4}$ of the circle.

2. Shade $\frac{2}{3}$ of the hexagon.

3. Shade $\frac{5}{8}$ of the square.

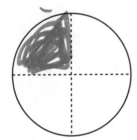 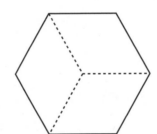 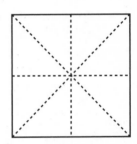

4. Circle the names of the shapes for which you shaded more than $\frac{1}{2}$ of the shape.

circle   hexagon   square

5. Divide the rectangle into fourths.

Shade $\frac{1}{2}$ of the rectangle.

How many fourths did you shade? _____

Practice

Solve.

6. 5 hundreds, 6 tens, and 9 ones = _____

7. 7 hundreds, 4 ones, and 3 tens = _____

225

Unit 10: Family Letter

End-of-Year Reviews and Assessments

In Unit 10, children will review the concepts and skills they have learned throughout the year. Children will review ways to make sense of collections of data, such as height measurements. Specifically, they will use the data they collected at the beginning of the year to determine how much they have grown during the last few months.

Children will also continue to use mental math strategies to solve number stories involving money.

Finally, children will review the following skills:

◆ Telling time to 5 minutes on an analog clock

◆ Using straws to construct geometric figures

◆ Reading and comparing temperatures on a thermometer

◆ Understanding place-value concepts

◆ Using the number grid

227

Do-Anytime Activities

To work with your child on the concepts reviewed in this unit, try these interesting and rewarding activities:

1. Continue to work on telling time to the minute.

2. Ask for answers to number stories that involve two or more items. For example, I want to buy a bran muffin for 45 cents and a juice box for 89 cents. How much money do I need? ($1.34) Encourage your child to use mental math, coins, the number line, or the number grid to work out solutions.

3. Point to a 3-digit number, such as 528. Ask what the digit "2" means (20); the "5" (500); the "8" (8).

4. Have your child create the largest and smallest numbers given 2 or 3 digits.

5. Together, note the temperature when the weather feels too hot, too cold, or about right. Encourage your child to read any temperature sign or billboard when you travel, noting whether the degrees are Celsius or Fahrenheit.

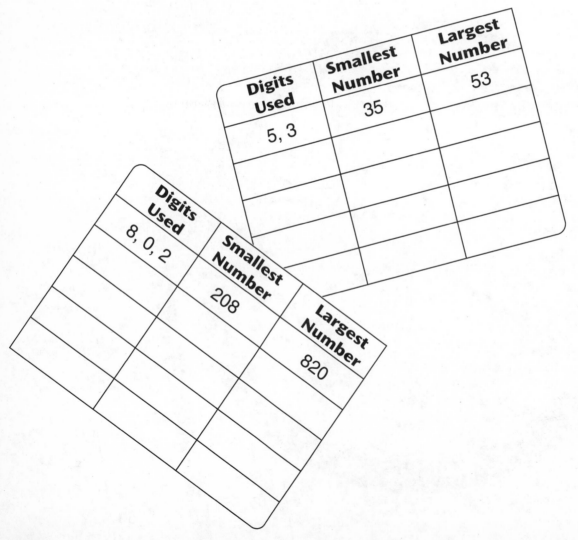

Digits Used	Smallest Number	Largest Number
5, 3	35	53

Digits Used	Smallest Number	Largest Number
8, 0, 2	208	820

Building Skills through Games

In this unit, your child will practice addition and money skills by playing the following games:

Beat the Calculator

See *My Reference Book,* pages 124 and 125. A "Calculator" (a player who uses a calculator to solve the problem) and a "Brain" (a player who solves the problem without a calculator) race to see who will be first to solve addition problems.

$1, $10, $100 Game

Players roll a die and put that number of dollars on their mats. Whenever possible, they exchange 10 dollars for a $10 bill. The first player to make an exchange for a $100 bill wins!

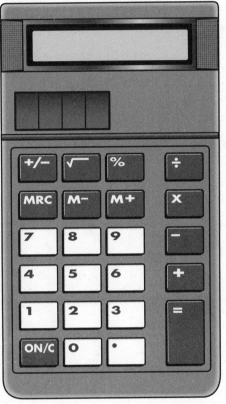

As You Help Your Child with Homework

As your child brings assignments home, you may want to go over the instructions together, clarifying them as necessary. The answers listed below will guide you through the Home Links in this unit.

Home Link 10·1

1.–4. Check that your child correctly graphs the birth months of families and friends, correctly identifies the months with the greatest and fewest number of births, and can tell you the number of births in those months.

5. >   **6.** =   **7.** >

Home Link 10·2

1.–2. Check that your child sets the hands on the clock correctly to show the times given; and that your child writes and says the times shown on the clock correctly.

3. 15   **4.** 30

5.

$\frac{1}{10}$	$\frac{1}{10}$	$\frac{1}{10}$	$\frac{1}{10}$	$\frac{1}{10}$
$\frac{1}{10}$	$\frac{1}{10}$	$\frac{1}{10}$	$\frac{1}{10}$	$\frac{1}{10}$

Home Link 10·3

1. Sample answer: ⓆⓆⓆⒹⒹ
2. Sample answers: ⓆⓆⓆⓆ or ＄1
3. 6   **4.** 13   **5.** 20

Home Link 10·4

1. 15¢

2. 20¢

3. 90¢; 10¢

4. 44, 48, 52, 55, 57, 64

Home Link 10·5

1. pentagon

2. rectangle

3. octagon

4. hexagon

5. square

6. triangle

7. Sample answers: 735; 1,711; 20,703; 799

Home Link 10·6

1. 48   **2.** 72   **3.** 46

4. 58   **5.** 80   **6.** 24

7. 105; 100; 95; 90

Home Link 10·7

1. 325; 334; 335; 346; 347; 355; 356

2. 704; 706; 707; 714; 715; 717; 727

3. 558; 568; 576; 578; 585; 586; 587

4. 931; 942; 943; 950; 951; 952; 962

5. 4   **6.** 4   **7.** 7

HOME LINK **10·1** **Graphing Birth Months**

Birth Months of Friends and Family Members

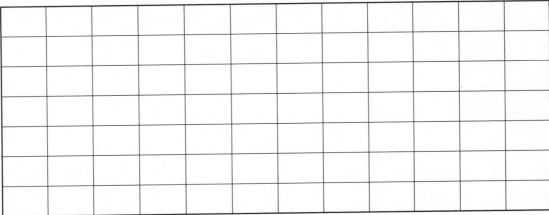

Jan. Feb. Mar. Apr. May June July Aug. Sept. Oct. Nov. Dec.

1. Which month had more births than any other month? _____

2. How many births were in that month? _____

3. Which month had the fewest births? _____

4. How many births were in that month? _____

Practice

Write <, >, or =.

5. 40 ____ 36

6. 123 ____ 100 + 23

7. Ⓠ Ⓠ Ⓠ ____ Ⓠ Ⓓ Ⓝ Ⓝ

Telltng Time

HOME LINK
10·2

Family Note Today we reviewed how to tell time to the nearest half-hour, quarter-hour, and five minutes. We also set clocks to a given time and then counted the minutes to a later time.

Help your child answer the questions below. Use the paper clock your child brought home earlier this year or use a watch or clock on which you can easily see the minute marks and move the hands.

Please return this Home Link to school tomorrow.

Have someone at home help you find a clock or watch that you can use to set the hands to practice telling time.

1. Ask that person to tell you a time. Set the hands of the clock to show the time. Do this a few more times.

2. Ask the person to show a time on the clock. Say the time and write it the way it looks on a digital clock. Do this a few more times.

Try these problems.

3. Set the clock to 2 o'clock.

How many minutes until quarter-past 2? _____ minutes

4. Set the clock to 4:15.

How many minutes until quarter-to 5? _____ minutes

Practice

5. Label each part with a fraction.

Color $\frac{2}{10}$.

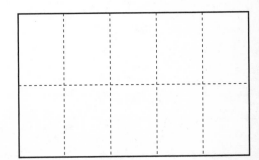

Solving Number Stories

Family Note Ask your child to explain what he or she did to solve Problems 1 and 2 below. Your child may want to model the problems with coins.

Please return this Home Link to school tomorrow.

For each problem, use Ⓟ, Ⓝ, Ⓓ, Ⓠ, or $1
to show the amount you pay.

1. You want to buy a ✏️ and a 🚗.

How much will you pay? _____

2. You want to buy a ⭕ and a 🍫.

How much will you pay?
Show the amount in two different ways.

Practice

Write the missing numbers.

3. 12 = 6 + _____   **4.** 7 + _____ = 20   **5.** _____ + 14 = 34

235

HOME LINK 10·4 Comparing Costs

Family Note Ask your child to explain how he or she solved the problems on this page. Encourage your child to act out the problems with coins or draw pictures of base-10 blocks.

Please return this Home Link to school tomorrow.

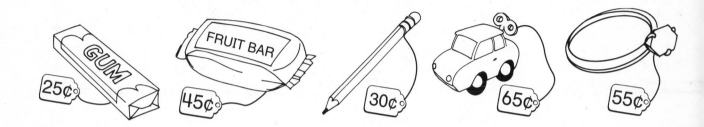

1. A [FRUIT BAR] costs how much more than a [pencil] ? _____ ¢

2. You buy a [ring]. You pay with Q Q Q.

How much change will you get? _____ ¢

3. You buy a [car] and [GUM].

How much will you pay in all? _____ ¢

You pay with $1. How much change will you get? _____ ¢

Practice

4. Complete the number-grid puzzle.

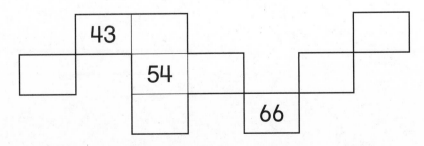

Name _____   Date _____

Geometry Review

> **Family Note** Today we reviewed several ideas about polygons and 3-dimensional shapes. Ask your child to point out objects of various shapes around the house or outside.
>
> *Please return this Home Link to school tomorrow.*

Use the Word List for Exercises 1–6.

Word List		
hexagon	octagon	pentagon
rectangle	square	triangle

Write the name under each shape.

1.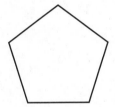

_____

2.

_____

3.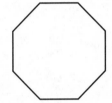

_____

4. I have 6 sides and 6 corners. What am I? _____

5. I am a special rectangle. All of my sides are the same length.

What am I? _____

6. I have the smallest number of corners of all of the shapes.

What am I? _____

Practice

7. Write 4 odd numbers with 7 in the hundreds place.

_____   _____   _____   _____

239

HOME LINK
10·6

Comparing Temperatures

Family Note

The focus of this Home Link is on finding how much warmer or colder one temperature is than another. Ask your child to explain how he or she solved each problem. One strategy might be to count on the thermometer or on a number grid. Your child might be able to solve some of the problems mentally.

Please return this Home Link to school tomorrow.

1. Which temperature is
   10° warmer than 38°F? _____°F

2. Which temperature is
   20° warmer than 52°F? _____°F

3. Which temperature is
   40° colder than 86°F? _____°F

4. Which temperature is
   20° colder than 78°F? _____°F

5. Which temperature is
   30° warmer than 50°F? _____°F

6. Which temperature is
   20° colder than 44°F? _____°F

Practice

7.

Rule
−5

110				

More Number-Grid Puzzles

Family Note Today we reviewed place value for 2-digit numbers such as 35, 3-digit numbers such as 827, and 4-digit numbers such as 1,254. We also completed number-grid puzzles for 3-digit numbers. Ask your child to explain how he or she solved each problem below.

Please return this Home Link to school tomorrow.

Fill in the missing numbers.

1.

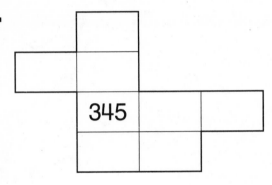

345

2.

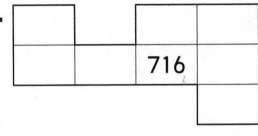

716

3.

588

4.

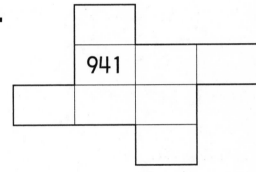

941

Practice

Solve.

5. 4 + _____ = 8   **6.** 10 = 6 + _____   **7.** _____ = 8 − 1

Family Letter

End-of-Year Family Letter

Congratulations! By completing *First Grade Everyday Mathematics* your child has accomplished a great deal. Thank you for all of your support!

This Family Letter is provided for you to use as a resource throughout your child's school vacation. It includes a list of Do-Anytime Activities, game directions, an Addition/Subtraction Facts Table, and a sneak preview of what your child will be learning in *Second Grade Everyday Mathematics.*

Enjoy your summer!

Do-Anytime Activities

To help your child review many of the things he or she has learned in first grade, we suggest the following activities for you and your child to do together over the summer. These activities build on the skills your child learned this year and help prepare him or her for *Second Grade Everyday Mathematics.*

Telling Time and Using Money

◆ Practice telling time by using a variety of clocks—billboard clocks, wristwatches, clocks with hands, and digital clocks—in a variety of situations.

◆ Set alarm clocks and timers on objects such as ovens, microwave ovens, and DVD players.

◆ Record the time spent doing various activities.

◆ Use real money in a variety of situations: allowance, savings, purchases (including getting change back), and using vending machines.

Weather Watch

◆ Invite your child to share your interest in weather predictions and temperature reports from the radio, the television, and local and national newspapers.

◆ Observe temperatures shown on business signs, aquarium thermometers, and so on.

◆ Read and set temperatures on heating and cooling thermostats and oven dials.

Beginning Geometry

◆ Look for geometric shapes in the real world, such as street signs, boxes, cans, construction cones, and so on.

◆ Construct polygons (2-dimensional shapes) using drinking straws and twist-ties from plastic storage bags. Small-diameter straws are easier to work with and are easily cut into 4-inch or 6-inch lengths. If only large-diameter straws are available, fold back the ends of the twist ties for a tighter fit. To build the polygons, put two twist-ties (or one folded twist-tie) into one end of each straw so that each end can be connected to two other straws.

◆ Construct 3-dimensional figures using straws and twist-ties. (It helps to connect the base straws first.)

Continuing with Scrolls and Number-Grid Puzzles

◆ Have your child fill in blank number grids and tape them together in order. This will help your child see two basic patterns of our base-ten numeration system:

1. You can write any positive number by using one or more of the digits 0 through 9.

2. There is no end to counting numbers—there is always at least one more, no matter how far you count.

◆ Here are two problem-solving challenges:

1. Have your child fill in the cells on a piece of a number grid to create letters of the alphabet, patterns, and designs.

2. Create puzzles from pieces of number grids in which most of the numbers are missing.

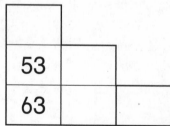

Fact Power and Games

Knowing basic addition and subtraction facts is as important in learning mathematics as knowing words by sight is in learning to read. Games are a fun way to provide the frequent practice children need in order to gain fact power, or the ability to automatically recall basic addition and subtraction facts. Children will build on their fact power in second grade, especially as they move on to computational skills with multidigit numbers.

The following section lists directions for games that can be played at home. The number cards used in some games can be made from 3" by 5" index cards or taken from a regular deck of playing cards. Cutout dominoes can also be used in place of number cards.

Addition Top-It

Materials ☐ number cards 0–20
(2 sets)

Players 2 or more

Directions

Players combine and shuffle their cards and place them in a pile, facedown. Each player takes 2 cards from the top of the pile and says the sum of the numbers. The player with the greater sum takes all of the cards then in play. The player with the most cards is the winner. Ties are broken by drawing again—winner takes all.

Beat the Calculator

Materials ☐ number cards 0–10
(4 of each)

☐ calculator

Players 3 (a Caller, a Calculator, and a Brain)

Directions

Shuffle the cards and place the deck facedown. The Caller turns over the top 2 cards. The Calculator finds the sum of the numbers on the cards by using a calculator. The Brain solves the problem without a calculator. The Caller determines who got the correct answer first. Players trade roles.

Penny Grab

Materials ☐ 20 or more pennies; paper and pencil

Players 2 or more

Directions

Each player grabs a handful of pennies, counts them, and records the amount with cents and dollars-and-cents notation. For example, a player would record 13 pennies as both 13¢ and $0.13. Partners compare their amounts and then figure out and record how many in all (the sum). Players repeat the grabs several times.

Variation: Use nickels or dimes.

High Roller

Materials ☐ 2 dice

Players 2 or more

Directions

One player rolls 2 dice. The player keeps the die with the larger number (the High Roller) and throws the other die again. The player then counts on from the number rolled on the first die to get the sum of the 2 dice.

Your child can also practice addition and subtraction facts on the Addition/Subtraction Facts Table. You can use this table to keep a record of facts that your child has learned.

+, −	0	1	2	3	4	5	6	7	8	9
0	0	1	2	3	4	5	6	7	8	9
1	1	2	3	4	5	6	7	8	9	10
2	2	3	4	5	6	7	8	9	10	11
3	3	4	5	6	7	8	9	10	11	12
4	4	5	6	7	8	9	10	11	12	13
5	5	6	7	8	9	10	11	12	13	14
6	6	7	8	9	10	11	12	13	14	15
7	7	8	9	10	11	12	13	14	15	16
8	8	9	10	11	12	13	14	15	16	17
9	9	10	11	12	13	14	15	16	17	18

Looking Ahead: *Second Grade Everyday Mathematics*

Next year, your child will …

◆ explore multiplication and division.

◆ use arrays, diagrams, and pictures to solve multiplication and division number stories.

◆ read and write 5-digit numbers.

◆ compare fractions.

◆ find the range and median of a set of data.

◆ classify 2- and 3-dimensional shapes.

◆ use tools to measure length, area, weight, capacity, and volume.

Again, thank you for all of your support this year. Have fun continuing your child's mathematics experiences throughout the summer!